The Ocean World of Jacques Cousteau

The Act of Life

The Ocean World of Jacques Cousteau

Volume 2
The Act of Life

THE DANBURY PRESS

The sea is wide and deep. A small animal might travel through it a lifetime without ever finding a mate. This is certainly one of the reasons why many species of fish happen to "school." In addition, bunched up in close-order array, on the constant lookout for food and foe, the schooling fishes have maximized their protection from predators.

The Danbury Press
A Division of Grolier Enterprises Inc.

Publisher: Robert B. Clarke

Published by Harry N. Abrams, Inc.

Published exclusively in Canada by
Prentice-Hall of Canada, Ltd.

Revised edition—1975

Project Director: Peter V. Ritner

Managing Editor: Steven Schepp
Assistant Managing Editor: Ruth Dugan
Senior Editors: Donald Dreves
                Richard Vahan
Assistant Editor: Sherry Knox

Creative Director and Designer: Milton Charles

Assistant to Creative Director: Gail Ash
Illustrations Editor: Howard Koslow

Production Manager: Bernard Kass

Science Consultant: Richard C. Murphy

Printed in the United States of America

234567899876

LIBRARY OF CONGRESS CATALOGING
    IN PUBLICATION DATA

Cousteau, Jacques Yves.
    The act of life.

    (His The ocean world of Jacques Cousteau;
v. 2)
    1. Marine fauna.   2.   Reproduction.
I. Title.
[QL122.C63 1975]   591.1'6   74-23065
ISBN 0-8109-0576-0

Contents

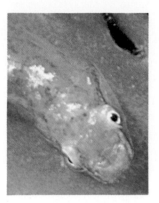

Sexual reproduction is carried on in any number of ways—and does not necessarily imply physical contact between the sexes, as in the case of species where eggs are released wholesale into the sea to be fertilized by any passing male. But in the forms of PEACEFUL ROMANCE (Chapter VII) we take a look at various means by which animals of the sea seek the attention of potential mates.

All is not peaceful romance in the ocean. There is plenty of VIOLENCE FOR SEX (Chapter VIII). From pygmies like the Siamese fightingfish, through lethargic monsters like the grouper, to the great seal family the male must find a mate and find a territory where the pair can get down to the business of creating the new generation.

Romance, hot or cool, has a single goal: the production of new life in sufficient quantities that species can meet their environmental challenges. Much of this new life can be called CHAN-CELINGS (Chapter IX), the billions of eggs or larvae that are simply dumped into the ocean to make their own ways in the world.

With those species that practice INTERNAL PARENTAL PROTECTION (Chapter X) the animal world has taken a big step in forwarding the survival chances of the young.

Beyond this, the animal world has taken another step forward by providing PARENTAL CARE FOR THE YOUNG (Chapter XI) after birth. In this activity males often participate—as with seahorses and the mouth-breeding catfish, or that very busy nest-builder, the stickleback.

In the sea as on land no population of animals is ever entirely free of PERVERSIONS (Chapter XII). When animals cannot move or respond to their basic drives in the ways demanded by instincts rooted deep in their beings, and perhaps millions of years old—in captivity, for example, or in conditions of extreme crowding—their behavior loses its coherence.

Scientifically we may not be able to define the word, but the fact remains that the BEAUTY (Chapter XIII) of the act of life in the sea is overwhelming.

Finally, in the sea THE ACT OF LIFE is also the act of love. In the sea procreation is almost always more than an anatomical expression. Rather it is a body of related activities bathed in a special ambiance of its own—with innumerable rites and rituals running from "love at first sight" to nursing and teaching the young in families.

Introduction: Dying for Survival

Imagine an immortal animal. Evolution so fashioned his glands and organs that his parts replace themselves as soon as they cease to function. His teeth may wear down or be knocked out, but he always has another set handy. His joints never suffer from arthritis; his legs remain forever as springy as an adolescent's. His bones do not grow brittle or his skin flabby. Cataract is unknown to him. No plaques of cholesterol deposit on his arterial walls. His heart muscle and the alveoli of his lungs renew themselves. He is invulnerable to cancer and all forms of viral, microbial, bacterial infection. He exists in total harmony with his environment—never too cold or hot, never hungry or thirsty, never wanting for oxygen. As he has no reason to die he doesn't, but lives on and on through the ages—growing a bit bored, perhaps, but animals seem to agree that life is better than death on almost any terms.

Has evolution ever produced such a prepossessing creature? Theoretically it should be possible for the various components of the endocrine system of an animal so efficiently to collaborate that physical obsolescence is simply banished from its life program. We know of certain plants—for example, lichens and the bristlecone pines of California's Inyo Forest—that live many thousands of years, near enough to immortality so far as animal lifespans are concerned. But the oldest animal of which there is a record seems to be a tortoise that managed to struggle through 150 years (plus, perhaps, another 25 years or so)—not all that much older than many old men.

Paradoxically, if immortality has ever been attained, it has quickly been eliminated, simply because immortality cannot survive. An immortal animal would be a dead animal—the representative of a vanished species. For earth is constantly changing, and animals must be ready to adapt to earth's changes. As we saw in the first volume of this series, *Oasis in Space,* at least four times in the past 600 million years the reef communities around the world have been all but obliterated by upheavals in the environment still not completely understood. Skeletons of palm trees have been discovered in Antarctica. At one time there were meadowlands on that continent, now under hundreds of feet of ice, not unlike the plains of the American West. Faced with this dimension of drastic environmental transformation, any immortal animal would be helpless. His ideal adjustment to the old environment spells certain extinction in the new. Locked into his "perfection," he cannot adjust. Immortal or not, he must die.

The mechanism by means of which the animal world responds to the challenges of a changing environment, and by means of which species establish themselves and adapt and survive, is the births and deaths of individual animals. There is no such thing as a perfect animal, much less an immortal one. But in any large population there is one individual with a thicker hide, another with a more flexible snout, another with a bigger cerebrum, another with the tendency to bear twins, another with acuter hearing. And so on. In other words, any successful species presents the environment not with an army of perfect individuals but with a smorgasbord of different characteristics dispersed through its membership. Then, when the environment challenges the species, the species has a chance to come up with the answer. If the weather grows colder, for example, the thick-skinned individuals will tend to make it, the thin-skinned ones to die out. As the cycles of sex, birth, and death follow one another down the generations eventually all members of the species carry the thicker skin and there is an adjusted balance between the demands of the environment and the capacities of the animal. The species has adapted and survived.

The process does not always work so simply. Immense as the dinosaur population was at the end of the Cretaceous period, some 65 million years ago, the smorgasbord of natural variations within the species was too limited for the challenges the environment posed it. The dinosaurs had gone too far down one particular evolutionary road; unable as individuals and species to find a solution to the radically altered climate, they died out. And there are other factors which compromise or exaggerate the "normal" operations of natural selection. These belong to the story of evolution. But man is the great zoological eccentric who should command our attention here. For man's brain gives him tools with which to participate directly, consciously, in his own evolution. To start with, he has grown as interested, or almost as interested, in individual survival as in species survival. Egged on by this concern for individual survival, he has learned to screen himself from many of the "natural" agents of the selection process. Puerperal fever, tuberculosis, pneumonia, smallpox, diphtheria, plague— all these grim reapers which for thousands of years winnowed out the human species are mostly fears of the past. A boy who 150 years ago might have died of whooping cough can now live to maturity—perhaps to become a great physician who discovers the cure for yet another man-killing disease.

If man is a unique exception as an animal it is a recent phenomenon. He is what he is because of the loves and deaths and births of uncountable legions of animals who have lived and perished since the first tiny cells stirred in the ancient oceans 3 billion or so years ago. From these lowly entities the natural-selection process has moved man steadily forward: past the jellyfish and the mollusc, past the turnoffs to equally successful evolutionary strategies like the insects', into the early experimental chordates, the mammals with their invaluable specialty of caring for their young, to the primates—to *himself!* Trillions of generations, trillions of deaths—each one a small link in the chain of evolution, each one a survival-ticket for man. Biological sciences are only a few generations short of being able to interfere consciously with genetics and to produce eternal youth. If we are reasonable enough to avoid a nuclear holocaust and to control population, immortality will no more be a utopian dream.

Jacques-Yves Cousteau

Chapter I. The Fecundity of the Sea

From the vast expanses of its surface waters to its beaches and marshes and tidelands and mangrove swamps, from its many thousands of miles of rocky shores to its deepest and darkest abyss, the sea produces life in fantastic abundance.

No wonder. The oceans are superior to land as an environment for life support. They provide directly the water fundamental to all forms of growth, laden with the salts, dissolved gases, and minerals.

The oceans also provide more constant temperatures, reliably warmer in shallow and surface areas, reliably cooler in their deeps —freeing many species from the need to adapt, as most land animals must, to wide variations in temperature.

The ocean's buoyancy also provides the most desirable home for most varieties of eggs, the great majority of which float near the surface waters where sunlight and oxygen help to develop them to larval stages, and where currents can carry them to nursing grounds in time for hatching. It is interesting to find that the eggs of most freshwater fish are heavier than those of marine species. In most cases freshwater eggs drop to the bottom after spawning, so that similar

> "Only by immense fertility can many species survive, and even the barest survival is absolute success."

currents will not carry them into the oceans and out of their natural habitat.

Only by immense fertility can many species survive, and even the barest survival is absolute success. Many marine creatures, unable to protect their eggs after spawning, have de-

veloped the capacity to produce enormous quantities at a time. In this way, a few of the eggs are given better odds for survival against predators and other controlling factors. The blue crab will produce several thousand eggs at a time. The average mackerel will produce as many as 100,000. This

> "The average mackerel will produce as many as 100,000 eggs at a time. This is nothing. A hake will produce perhaps 1 million eggs at a time, and a cod from 2 to 9 million. In a year an oyster will lay 500 million eggs."

is nothing. A hake will produce perhaps 1 million eggs at a time. A haddock will lay anywhere from 12 thousand to 3 million, and a cod from 2 million to 9 million. The purple sea hare will produce 20 million eggs, and the mola mola as many as 28 million. In a year, an oyster will lay 500 million eggs!

Female North American lobster. In this extraordinary photograph, we see the abdominal surface of the female North American lobster and a few of the 3000-100,000 eggs she lays every two years. The eggs remain attached to the mother's underside and tail for about ten months. Naturally, huge numbers of them are rubbed off, to be nipped up by predatory fishes, as the female walks about. From those which survive into the hatching period emerge free-swimming larvae—looking nothing at all like an adult lobster. The larval stage lasts through five or six molts taking six to eight weeks. The few larvae that manage to make it through this period finally sink to the bottom and hide themselves in the mud or among rocks. In their first year they will molt another 17 times or so; the young lobster will now be about five inches long. So far as is known there is no maximum size or age for a lobster; they just keep growing. The largest ever caught weighed 42 pounds. But down there on the bottom somewhere there's undoubtedly one that's much bigger.

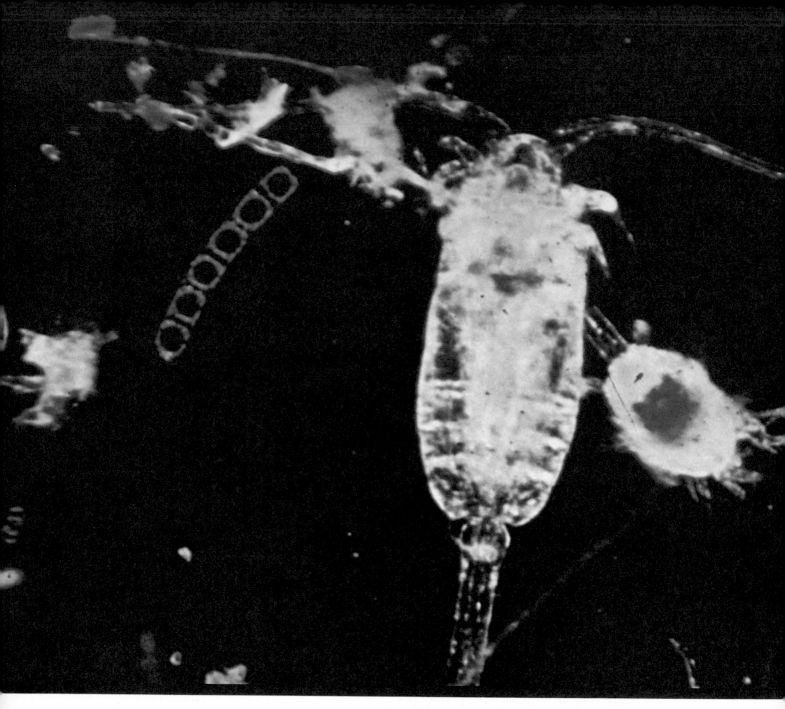

In the Broth of the Sea

The ocean is deceptive. There is far more in it than meets the naked eye. As much vegetation exists in the average acre of surface sea as in an acre of earth. In the average cubic foot of this water as many as 20,000 microscopic plants will be found, together with hundreds of planktonic animals. (The word *plankton* is from the Greek. It literally means "that which is made to wander.")

What slips through the finest net is even more impressive. The same cubic foot of water may hold well over 12 million plant cells, or diatoms—unicellular creatures which absorb the salts and minerals from the sea and, using the energy from the sun, photosynthesize the vitamins, starches, sugars, and proteins on which plankton animals feed. Diatoms comprise three-fifths of all planktonic life. Encased in its self-made crystalline housing of silicon, each cell re-

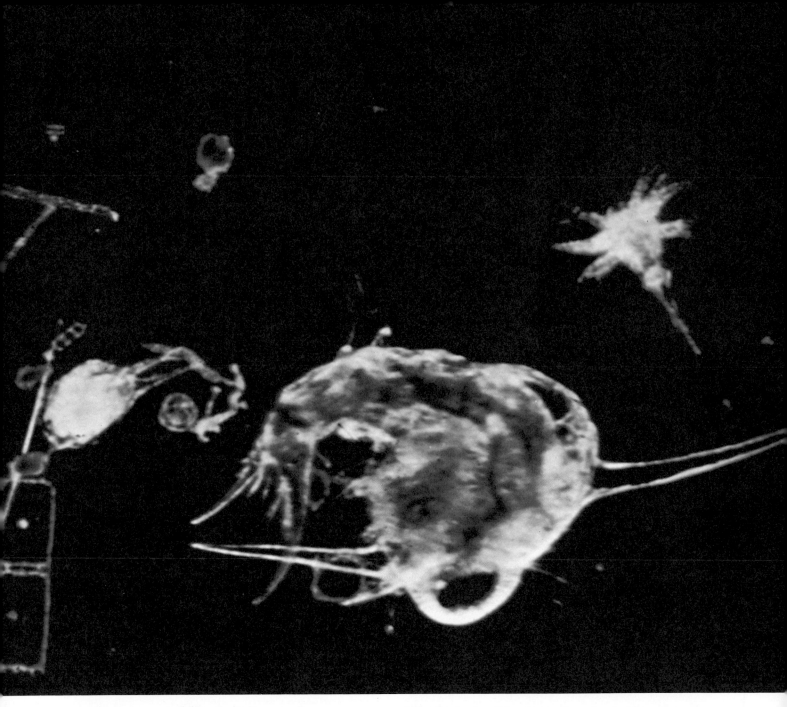

produces so rapidly that it may have more than 1 billion progeny in a month.

Riding along in this miniature jungle are foraminifera, radiolaria, salps, and tiny crustaceans. Of this last group the copepods, shrimplike creatures, many no larger than the head of a pin, not only outnumber any other kind of multicelled animal in the world, but may well outnumber all of them together. A copepod may consume 100,000 diatoms daily. The average herring, in turn,

Plankton. In this photomicrograph, the more geometric-patterned plankton are phytoplankton—that is, the drifting microscopic plants of the sea. The others are microscopic crustacean larvae and a copepod.

will consume from 60,000 to 70,000 copepods in the same length of time, and a whale may consume 5000 herring in a single meal. And so, beginning in the plankton, the pyramid of life builds. Think of the astronomical cost to build a whale!

13

Chapter II. Controlling Factors and the Balance of Nature

An estimated 100 million plaice spawn each year in the North Atlantic. Each female will lay as many as 350,000 eggs. If all these eggs hatched and grew to maturity, untouched

> "The success of a single species, if too great, can wreak havoc throughout a system. Many species have a mortality rate that is well over 99 percent."

by predators, unmoved by currents, unhurt by severe changes in temperature, within a few months the hosts of maturing plaice would devour every species of every smaller edible creature in the sea. The life-forms on which these smaller creatures had fed would in turn proliferate to the point where they could not find sufficient food for themselves.

The success of a single species, if too great, can wreak havoc throughout this system. Many species have a mortality rate that is well over 99 percent. Not all of the attrition is caused by predators in the food chain. The growth of each species is controlled by a number of other important factors: water temperature, surface temperature, the strength and random flow of ocean currents, which may sweep vast quantities of helpless eggs and larvae away from their food sources or wash them ashore where the sunlight and the air will shrivel them dry.

To compensate for these dangers, fish and other animals in the sea have developed a number of ways to begin life and continue it. The majority of ocean species enter the world as eggs, and the eggs of most species are buoyant and remain near the surface, where they continue to develop as larvae. They need the sunlight and the oxygen in

order to grow. This "holiday" in the plankton, while necessary, is dangerous. Immobile and exposed, the young are easy prey.

Fishes like the ling cod mate in the open sea and leave their fertilized eggs to fend for themselves. They can afford to. Each female lays millions of eggs at a time. Some of the eggs are bound to survive the ravages of predators and other controlling factors. Those animals which are less productive, like man and the sea mammals, carry their offspring internally, suckle them for months, and help them to survive to adulthood.

Some eggs are laid on the ocean floor, or in crevices of rock, and are heavy enough to remain on the bottom, though currents can still detach them to become the prey of bottom feeders. Other eggs have a sticky surface, and affix themselves to rocks and weeds and shells. There are species like the grunion that bury their eggs on sandy beaches, others like the salmon that travel long distances to freshwater spawning grounds comparatively free of predators. Still other eggs ride with their parents—some on the underside of the mother's body; others, with added security, in a brooding pouch. There are species that incubate their eggs within the female body, that nourish the embryos after they hatch. To sum it all up, in the sea sheer numbers are the greatest protection against species extinction.

Ling cod. Here shown in its newborn state, the ling cod is capable of changing its bright colors with amazing speed. These fish spawn in the winter and leave their eggs in whitish-pink groups. With big, canine-toothed mouths, they feed on squids, herring, flounders, and their own young. The female of the species weighs as much as 70 pounds, while the male rarely weighs more than 20 pounds.

Survival of the Few

Just six weeks ago, 102 miles south of this peaceful scene, and a good deal farther out to sea, over 300 million mackerel came together to spawn—as they do every year at this time. The females produced approximately 10,000 trillion eggs, of which 9100 trillion were successfully fertilized. The eggs rose to the surface in great masses. Ordinarily the current would slowly have carried them southwest, closer to shore, to ideal nursing grounds for the larvae—to hatch within two weeks.

But just as these particular new eggs began to carpet the surface a strong wind rose from the south, forerunner of a distant storm. A few hundred trillion eggs kept drifting on their normal course but the great majority were picked up by the shifting currents and moved rapidly northward. In their first few hours on the surface they fell prey to hundreds of thousands of arrow worms in the plankton. Day after day the arrow worms steadily devoured the eggs at the rate of several million a day, so that within two

> **"Whales sift the plankton, each taking in as many as 1 billion eggs in a swallow."**

weeks they had taken over 500 billion of them. Bacterial and fungal diseases also attacked—another 100 trillion eggs or more lost in the same amount of time. Medusae were floating everywhere in the plankton, impaling thousands of eggs on their many tentacles. Many billions more eggs were lost to them. The larger losses came from larger predators, including a school of herring, which approached the vast egg col-

16

ony from the east. Each herring gobbled up from 10,000 to 20,000 eggs while the school stayed with the eggs. Then another school appeared, then another. Several whales began to sift the plankton, each taking in as many as 1 billion eggs in a swallow, also taking in herring at the rate of 2000 or 3000 —and disposing of another couple of hundred million eggs—every few hours. Countless other varieties of fish appeared and consumed the offspring.

As the eggs moved farther north, the temperature dropped below the normal level for hatching and this slowed their development. Before they could hatch, the storm was upon them, scattering the egg clusters in every direction. Most were blown northward at an accelerated rate, where the temperature drop was so severe that prodigious numbers of the embryos died. Several hun-

Black skimmers are among the many predators that prey on fish, larval fishes, and other drifting animal forms in the sea and along its boundaries.

dred billion eggs were washed up on this beach by the storm. When the winds subsided and the sun appeared, these stranded eggs, glistening wet and still containing life, began to dry and die. Thousands of small birds appeared, scooping up hundreds of eggs apiece. What they did not pick clean from the sand the land crabs did. Over the next few days the remaining eggs, now reduced to about 300 trillion, began to hatch. The mortality rate increased. It had been about 10 percent per day in the first few days of larval life. It would grow to 45 percent per day nine weeks later. About 10 out of every million eggs would become adult fish.

17

Dying for Survival

As humans, our views about life and death are extremely individualistic and subjective. In the wild, death has a broader meaning. We have seen (pages 8-9) that death is essential for the survival of the species. Thus, the ironic theme: To live in order to die.

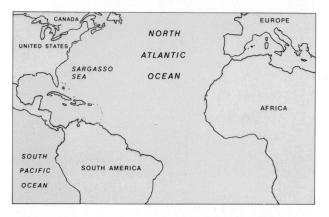

Many lives pursue singlemindedly the aim of preparing for and executing the act of life that ensures the next generation. The common eel (*Anguilla*) comes in two species—American and European. They are called freshwater eels, but this is a kind of misnomer: their entire freshwater existence is pointed at the Sargasso Sea, that legend-inspiring waste of warm, motionless waters and seaweed meadows created by winds and currents in the North Atlantic.

The European eel occurs in streams and rivers along the European west coast from Spain to Norway. In their freshwater habitat the eels take from five to 20 years to mature. When their time comes they undergo a fantastic metamorphosis. They change color from yellow to silver; their skin thickens; their eyes enlarge and alter form; they cease to feed. Their glandular system is deeply modified, which is necessary if the eel is to handle the salts of seawater. Armed with brand-new bodies they swim down their homestreams into the ocean—following currents nearly across the vast Atlantic

to the Sargasso Sea. Here they spawn and die—probably at great depths. Life in a cold Norwegian or Pyrennean freshet is but a prelude to this dramatic transformation into deep-sea fishes—ending in distant, dark ocean spaces.

The leaf-shaped larvae which hatch from the eggs laid in the Sargasso Sea drift to the surface and allow themselves to be swept along by the Gulf Stream back to the coasts of Europe—a journey taking about three years. Here the little animals encounter the first big change—into elvers.

The American eel found in the eastern part of North America has a similar life pattern —and oddly enough breeds and dies in the same Sargasso Sea as its European cousin, a little to the west and south of him. However,

> "When their time comes they undergo a fantastic metamorphosis. They change color from yellow to silver; their skin thickens, their eyes enlarge and alter form; they cease to feed."

it seems to take the American hatch only one year to get back to the mainland, to metamorphose into an elver, and to re-ascend a freshwater stream. In this species, upon arriving from the Sargasso Sea, the young males and females climb into fresh water on their own, and only rejoin at river mouth six years later for the death voyage to the spawning grounds.

The American common eel, which descends 1500 feet, lays as many as 10 million eggs, and dies. Their long tubular bodies average two to three feet.

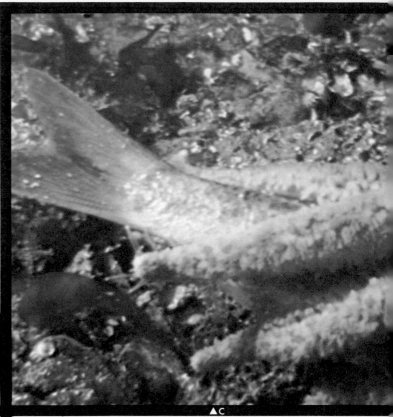

▲A   ▼B   ▲C

Survival Through Feeding

Those animals of the sea that have survived to this day have done so because they could adapt. The means by which they find and take their nourishment is one of the most important of these adaptations. Some developed methods of filtering their food from the water, while others learned to pounce on their prey. Still others learned physically to extend themselves in order to envelop their meals.

A / The gray shark is a fine example of a fish so well designed that it has changed little in millions of years. Streamlined, fluid swimmers, sharks are equipped with an extraordinary sense of smell. These bloodhounds of the sea have been observed to follow their prey in and out of coral outcroppings, round and round until the meal is captured and eaten. Sharks are known to bite at almost anything, often without sense; individuals have been found with blocks of wood in their stomachs.

▲D   ▼E

B / *Sea lions* and *seals* are animals whose ancestors once roamed the dry land. They have turned to the sea and find abundant food supplies there. Their diet consists of a wide variety of fish, crustaceans, squids, and sometimes seabirds.

C / *Sea stars* of the genus Solaster, differing from their five-armed relatives, may have as many as 24 arms surrounding a central disc. Possibly the largest starfish, this variety may attain a diameter of as much as two feet. Many sea stars feed primarily on slow-moving bivalves, but we see in this picture that the species can also feed on dead fish.

D / This rugged-looking *grouper* lives in reef crevices where it preys on the other fishes that call the reef home. A grouper may weigh from a few pounds to a few hundred pounds. Lethargic, he doesn't pursue his meals but instead waits until something inviting passes by. Notice the lack of large teeth; prey is sucked in when the mouth is opened.

E / The majestic *manta ray* flies like a giant eagle in its domain. Ten- or 11-foot wings sweep up and down, propelling this giant through concentrations of small marine life. The flaps adjacent to the large mouth direct the flow of food down its throat.

The Perfect Example

The coral reef is a remarkably interdependent community, with a greater abundance of life per square foot than anywhere else in the sea. The corals build their limestone skeletons from calcium dissolved in the sea. Wave action breaks off pieces, and these grind into sand on the shallow side of the coral, gradually forming a beach. Various kinds of algae grow over the floor and proliferate in the sunlight. Sea urchins cluster about the base of the coral, starfish and rays and sea cucumbers move across the sandy bottom. Tube worms open their parasols to share the planktonic rain, and shrimps and lobster and fierce moray eels fill the grottos and crevices. Giant tridacna clams live in the outer layers of the coral. Filefish eat the polyps. Parrotfish bite into the coral blocks and crush the limestone into sand. Hundreds of other varieties of fish gather at the reef to feed on this life. Thus, a coral reef is a perfect example of the balance of nature. The controlling factors adequately reduce the astronomical fecundity-potential of the reef community. And the greater the number of species, the greater the stability.

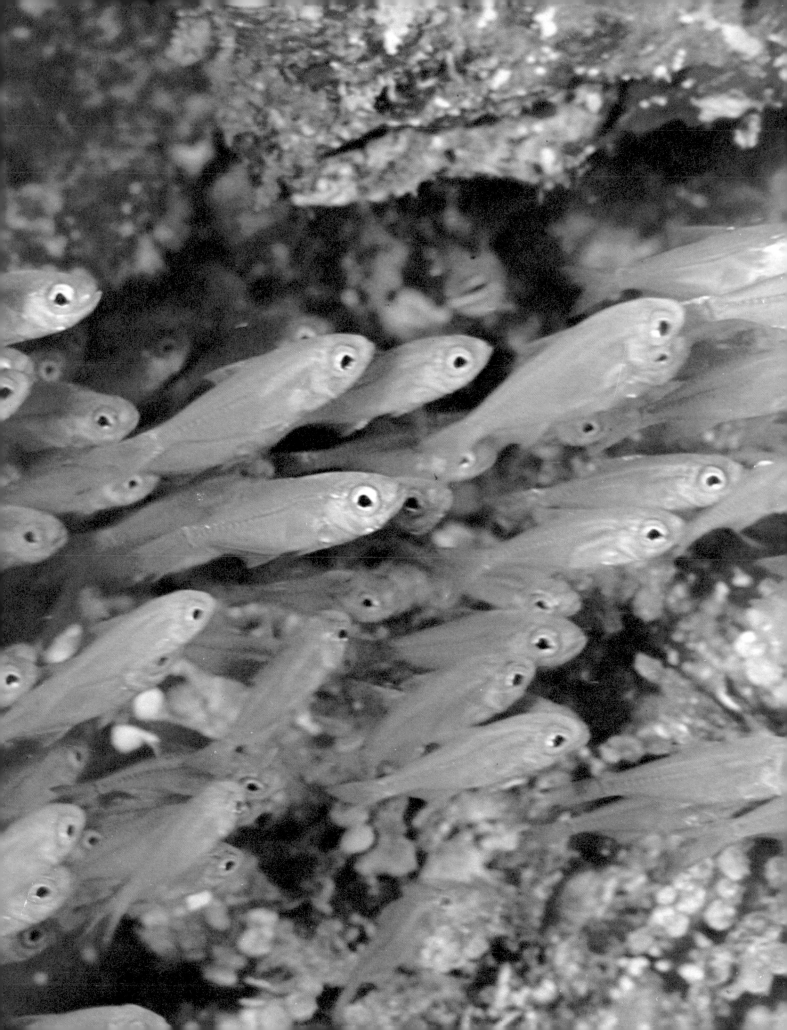

Chapter III. Survival of the Fittest

Four hundred million years ago fish had few fins and no jaws. They moved awkwardly along the ocean floor, sucking what food they could find from the mud.

Eventually some fish appeared with primitive jaws and with larger fins than others. This was accidental, but it gave them better steering and balance, more speed and the ability to catch other fish that had no jaws. They found more to eat. So a greater proportion of jawless fish with smaller fins, faring less successfully, died sooner, while a greater proportion of the new breeds survived longer. This latter group had the opportunity to reproduce oftener, contributing a greater number of offspring to each subsequent generation. Because, when they reproduced, they passed on their characteristic as genetic information, subsequent generations exhibited a larger proportion of faster fish.

Today, as a result of these processes, thousands of varieties of fish have developed, each species highly specialized for success in its environment. The parrotfish feeds on coral. It has developed a pair of jaws of solid hard bone, like a beak, plus a set of thick teeth fused internally. But there was a time when few parrotfish had beaks, few butterflyfish had spots on their tails; when flounder and sole swam vertically like other fish; when the viperfish did not have a stomach that stretched. All these special characteristics evolved by selection, in the course of which process the environment acted rather like a sieve to strain out individuals least suited to it, death in each generation reducing their number and increasing the majority of the fittest.

Some animals did not have to change. The great American comparative anatomist A. S. Romer writes, "The turtles, within the shelter of their armor, became the conservatives of the reptilian world. The oldest forms were contemporaries of the earliest dinosaurs. The dinosaurs passed away, and mammals took their place, but the turtles went calmly on their placid way. They learned to pull in their heads, but otherwise remained much the same. Now man dominates the scene, but the turtles are still with us. And if, in the far distant future, man in turn disappears from the earth, very likely there will still be found the turtle, plodding stolidly on down the corridor of time." This is species success—if ever there was.

The Galápagos tortoise you see opposite provides even more matter of interest to modern biologists. Over his bizarre, science-fiction island group in the Pacific he has proliferated—a separate species per island! That is food enough for thought. But these are oceanic islands far out from land. How did this heavily, hopelessly land-dwelling animal ever get to these isolated outposts in the first place? No one knows.

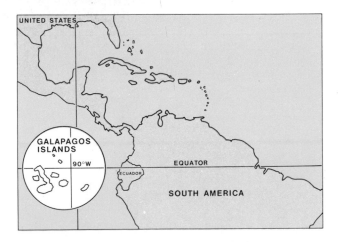

The Galápagos tortoise. This giant, 400-pound creature is strictly a land dweller, and in fact will sink instantly if placed in water. It is a cactus-eating vegetarian who gets his water supply through this food source. The average age of the Galápagos tortoise is estimated around 150 years!

The Evolution of Fishes

Fishes are one of five classes of the phylum Chordata, which also includes birds, reptiles, amphibians, and mammals—the class to which man belongs. These are the vertebrates, animals with backbones. Mammals are distinguished by their warm blood and hairy bodies. Birds also have high body temperatures and are set apart by their feathers and wings. Reptiles are a lower form of land life. The fourth class is that of the amphibians. The fifth, fishes, is that of the water dwellers with gills, animals that move by using fins rather than limbs.

Fishes originated more than 425 million years ago, in the Ordovician period. Fossils that tell of the transition from invertebrates are difficult to find, for primitive animals have few hard parts to fossilize. But about 400 million years ago (toward the end of the Silurian age), three distinct families of fish had developed. The earliest family was a group of fishes without jaws (Agnatha). They died out some 340 million years ago, although the lamprey eel, with its jawless mouth, might be a survivor of the primitive group with but the minor development of a ring of sharp teeth. Another group of fishes had developed primitive jaws and paired fins (Placodermi); their heads were covered with thick plates of bones; their jaw plates acted as teeth. This family died out 300 million years ago. The earliest family of spiny fishes (Acanthodii) lasted 50 million years longer, and from them developed two great families of higher bony fishes: lobe-finned fishes (Sarcopterygii) and the ancestors of today's fishes, the finned fishes (Actinopterygii). Though these also disappeared, two subfamilies, the coelacanths and the lungfishes, survive to this day.

The great weight of fish evolution in the next few million years took place among the ray-finned fishes, and most species which exist today developed from them (Teleostei).

While the early jawless fishes were dying out, and the bone-plated fishes were also on the wane, a new family suddenly began to flourish. These were the cartilaginous fishes (Chondrichthyes), including ratfishes, sharks, skates, and rays. They reverted to skeletons constructed of cartilage. Powerful and well-armed, they were remarkably suited to their environment and they still are today. Sharks and their brethren have

> "Sharks and their brethren have undergone little evolutionary change in the last 350 million years.
> They require none.
> Illustrating Darwin's principle that the fittest survive, these primitive species are more numerous today than ever in their long history."

undergone little evolutionary change in the last 350 million years. They require none. Illustrating Darwin's principle that the fittest survive, these primitive species are more numerous today than ever in their long history.

Subtle evolutionary changes are made with each new generation and are necessary if a species is to continue to exist in a world.

Fish family tree. Of the three groups of fishes shown in the family tree, two are still represented in the world's waters and the third, the lobe-finned fishes, seems to be reaching the end of the line—as fishes. It was from this group, scientists believe, that the first terrestrial animals came.

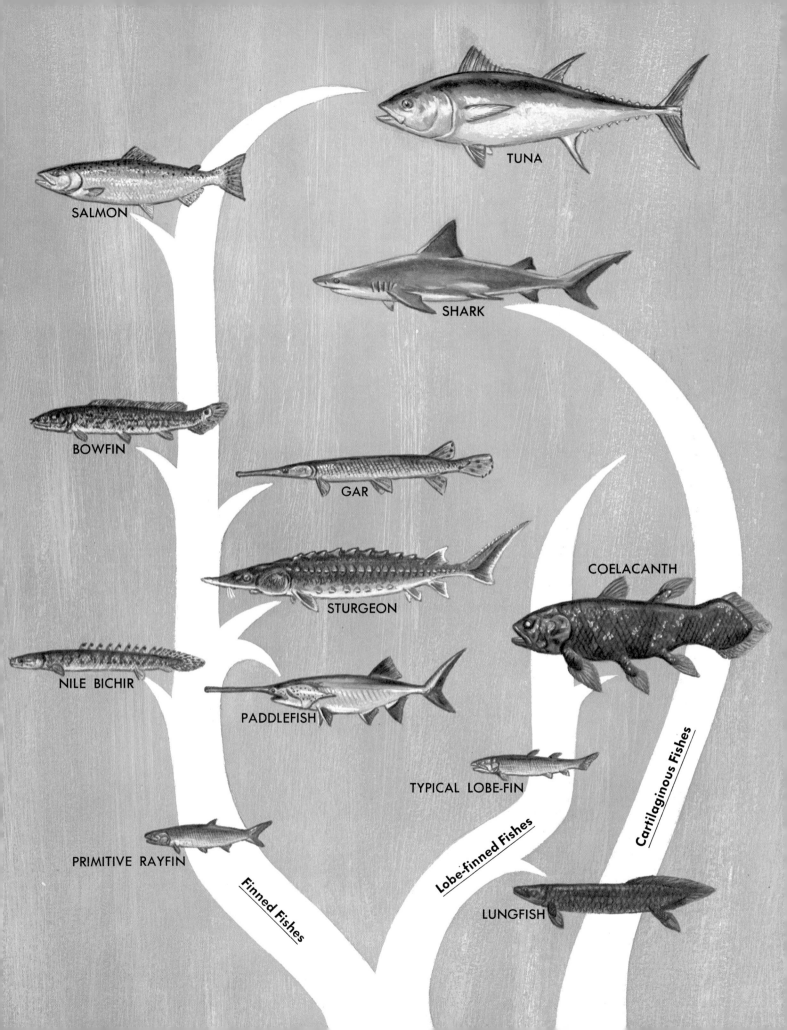

TUNA

SALMON

SHARK

BOWFIN

GAR

COELACANTH

STURGEON

NILE BICHIR

PADDLEFISH

TYPICAL LOBE-FIN

Cartilaginous Fishes

PRIMITIVE RAYFIN

Lobe-finned Fishes

Finned Fishes

LUNGFISH

Evolutionary Successes

A successful superior form in the sea—the dolphin: fragile, but intelligent, swift, loving, playful, candidate for man–dolphin intercommunication.

A successful inferior form in the sea—the gorgeous jellyfish, Pelagia, is a coelenterate. The digestive tract has only one opening—food is taken in and waste is ejected at the mouth.

Survival Through Death

The salmon pays for his success as a species with one of the most painful death agonies in the natural world. And only the superior individuals among the salmon could survive even as far as the death agony. In *Oasis in Space* I told this grisly tale. After four or five years in the open sea the North Pacific salmon heads for home. Guiding himself by

> "In a final furious frenzy the female digs a hole in the stream bottom, deposits her eggs, and the male fertilizes them.
> A few days later the pair die."

means of delicate odors dissolved in the water, or force fields generated by the ions of the sea salts as they move through earth's electromagnetic field, or by somehow employing the sun and stars as compass (these are theories as to how the salmon accomplishes his fabulous navigational feat), the mature fish presses on for the rivulet or pond where he began life. Swimming against the fierce flow of mountain streams, through rapids and over rocks and past an army of predators, terribly battered, fasting from the moment he embarks on this last trek, he forges on and on until he reaches the old spawning ground. Here in a final furious frenzy the female digs a hole in the stream bottom, deposits her eggs, and the male fertilizes them. A few days later the pair die.

In addition to the awesome demands nature makes of the salmon, scientists have been intrigued by the extremely rapid aging processes which are an aspect of this terminal phase of his life. In the final two weeks the salmon physically degenerates as much as a man would in 40 years: his arteries thicken, his liver gives out, his circulation weakens, he is subject to all kinds of infection and infestation. By the time he has spawned he is, in Dr. Andrew A. Benson's words, "a miserable shadow of the beautiful, silvery deep-ocean marine animal. His flesh has turned from orange-pink to pale tan. He has developed his hump and hooked jaw. His bones have become cartilaginous; his skin is peeling off. We even saw many with their tails falling off. His liver is a livid olive green because of the decomposition products of his hemoglobin. Only the heart of the salmon remains in good condition—and even this suffers from thickening of the coronary artery walls." Exhausted from malnutrition and the glandular ordeal, covered with fungus, the salmon has grown decrepit in a few days.

From the now white, blood-drained flesh and torn fins of the dying salmon physiologists hope to learn more about the aging process. Noting that salmon do not seem to suffer from heart attacks or strokes, researchers hope to apply to man clues obtained from these broken-down bodies. They hope to gain new insights into these disorders—as well as those of the liver, or arthritis, or even cancer.

Although the tale of the individual salmon is a sad one, it represents a triumph for the species. For in its death it provides for the survival of its kind. In the graveyard of the spawning grounds other fishes will feed on the battered bones of this marine phoenix. The decomposing body will nourish its own fry when they emerge from these "ashes."

Salmon saga. During the early part of their lives salmon live serenely in fresh water. The middle years are spent in the ocean, growing big and strong. When their time comes, they brave all sorts of dangers to get to the placidity of the old spawning grounds. They are here seen jumping rapids in order to reach these quiet waters.

Chapter IV. Treasure Map

Reproduction is essential for the continuation of the species, but how is it ensured that young salmon hatch from their eggs as salmon instead of baby ducks or alligators? The answer lies in a genetic set of instructions which reside within the nucleus of every cell. These instructions control every aspect of the lives of plants and animals. They control how the body is constructed; they direct the fabrication of enzymes which control all chemical processes; they determine the coloration of organisms and their behavioral characteristics. A remarkable amount of information is contained in each tiny cell nucleus.

Great advances in the understanding of genetics have been made in recent years, but there are still many aspects of this complex subject as yet unresolved. Around the turn of the century, research began to investigate the material within the nucleus of cells. A number of experiments showed that within this structure there was something that controlled the development of an embryo, and without the nucleus, development could not proceed. By using chemical stains and powerful microscopes, scientists found strands of material which appeared to thicken, contract, and then divide just before cell division. As the cell separated, one pair of each strand went to each new cell. These strands are known as chromosomes; they are condensed from chromatin material in the nucleus of the cell. Closer examination of them revealed numerous particles strung together like beads. Scientists theorized that these beads, subsequently known as genes, may be resposible for transmitting the hereditary traits from a parent cell to its progeny. Genes were named after the Greek word *genesis,* meaning "origin" or "birth."

By irradiating cells with powerful X rays, it was found that changes in the normal patterns of genes caused divided cells to develop abnormally. It was established that in some way genes work to control heredity. What was not established was *how* they worked.

It was time to look into the gene for the answers, to analyze its chemical elements. Nucleoproteins were found, molecules of a substance called deoxyribonucleic acid, or simply DNA. DNA consisted of deoxyribose

> "The DNA in all the chromosomes of one human cell contains more than 6 billion steps."

sugars, phosphates, and four nitrogen compounds: adenine, thymine, guanine, and cytosine—nothing else in the molecule.

The components of the gene were strikingly similar to that of the bacteriophage, a form of virus-infecting bacteria. The remarkable thing about the phage is that it attacks bacteria by entering each cell and taking over its machinery—the cell simply stops reproducing itself and begins to reproduce copies of its assailant. In effect, the phage has taken the raw materials of the cell and used them to create a new product.

A study of the phage disclosed that it was only DNA which entered the cell wall of bacteria. Now scientists knew that DNA was the active ingredient of heredity—the map, the "treasure map," we had been looking for. Somehow this substance creates a master plan, incredibly detailed, so that each gene knows its place on the ladder of the chromosome.

Links in the Chain

Here is the stuff of life. Constructed in 1953 by James Watson and Francis Crick, it is the first model of DNA to take account of all the information we now possess.

What you see is only a portion of a long chain of molecules containing the genes and making up the chromosome. The phosphates and deoxyribose sugars comprise the outer framework of the chromosome. The four nitrogen bases—adenine, thymine, guanine, and cytosine—link together in the middle. Each of the four bases may occur along either framework in any order. But whatever occurs on one side dictates precisely what may occur on the other. For each base can combine with one other, adenine only with thymine, guanine only with cytosine. Thus, if the links are parted, only one substance, identical to the original substance, can take its place to form a new link. Think of each link as a step in the genetic code. It is the order in which the pairs of linked bases occur that is now believed to dictate genetic character. There are infinite possible variations, just as many as there are forms of cellular life. There could be more than 1 billion combinations in every 15 pairs of links. A chain of links 1/1000th of an inch long could contain 85,000 steps. The DNA in all the chromosomes of one human cell would stretch for more than three feet and contain more than 6 billion steps—a very precise set of instructions indeed!

When the DNA molecules divide before cell reproduction, the links do in fact part in the middle, as a zipper might come undone. Additional units of the four nitrogen bases, accumulated by the cell and floating free in its nucleus, are now attracted to the exposed ends of either half of the DNA. They link up in the same order, creating two new DNA molecules that are replicas of the original.

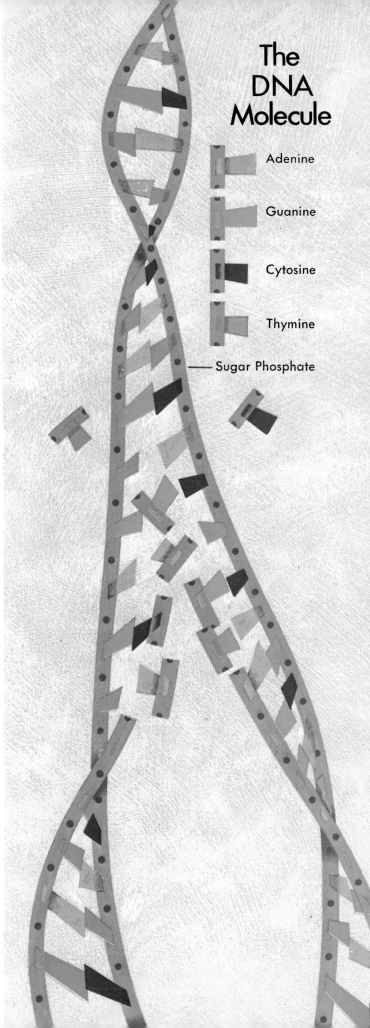

The DNA Molecule

Adenine

Guanine

Cytosine

Thymine

—— Sugar Phosphate

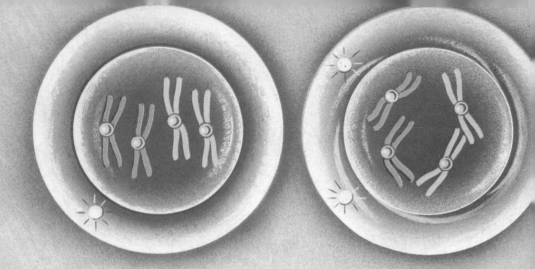

Mitosis

Meiosis

The Never-ending Process

Marine animals and plants grow by reproducing their body, or somatic, cells. This is accomplished by a division of the nucleus of each cell. The process is a continuous one, rapid in early stages of growth, enhanced by appropriate diet and ideal environmental conditions, somewhat hindered by aging and disease, but unending. It is called mitosis. Many one-celled planktonic plants and animals also reproduce by this method.

The several stages of its development begin when the chromatin thickens in the nucleus of the cell. Chromatin contains the genes that express any particular character in the species, from size, shape, coloration, weight, and the structure of internal systems—down to the most minute detail. The chromatin material forms into thin rods called chromosomes. Each chromosome divides lengthwise into pairs, after which there is twice the normal number of chromosomes. The cell next squeezes itself into a spindly shape, with half the chromosomes rushing to one end of the cell, half to the other. Then the spindle parts in the middle. The group of chromosomes caught at either end, each equal in number to the undivided chromosomes in the original cell, now dissolve again into chromatin, and new nuclei are formed. We now have two identical cells.

Sexual reproduction, or meiosis, takes place in a similar way to body growth but with

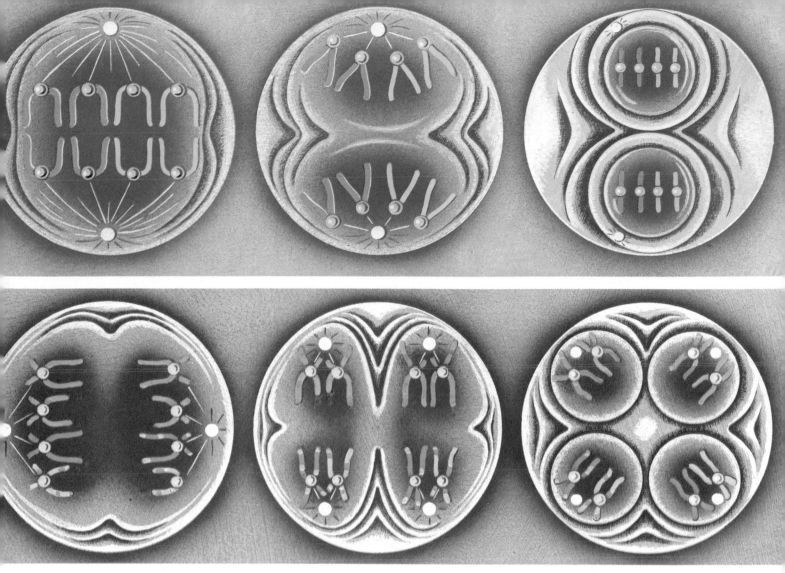

an important difference. Genetic variability must be ensured so that the character of the strongest and most successful members of the species may have the maximum opportunity to express itself, giving the species

> "Genetic variability must be ensured so that the character of the strongest and most successful members of the species has maximum opportunity to express itself."

optimum chances for survival. Cells identical in genetic character must be avoided as new generations are spawned. Thus, the sex cells, or gametes, in plants and animals, including both the female ovum and the male sperm, share the ancestry of each new and fertilized cell they produce. Each sex produces a haploid cell: that is, a cell with only half the required number of chromosomes for a normal (diploid) cell. To attain the full number they must unite, and this process of union completes fertilization. To come up with such haploid cells the gametes begin their cell division in the same way. Chromosomes divide, doubling in number, and then the cells divide. But now, before the chromosomes even have a chance, it would seem, to divide again, the cells divide a second time, creating four new cells, each with only half the needed number of chromosomes. After the union of male and female cells, the new individual develops further by asexual divisions of the fertilized egg.

The Making of a Shark

To imagine the information transmitted to this shark embryo to bring it to its present state of development we need only remember that it began within the female's body as a single cell no larger than the head of a pin. Specific directions carried forth by genes dictated that the egg divide again and again. Soon, as a new set of directions was issued, this simple division ceased. Now cells in one embryonic region were to follow a course resulting in the development of the adult ani-

> "Literally billions of predetermined decisions have been made to bring this embryo to its present state of development."

mal's brain. Another area of the developing embryo was directed to form the digestive

system and its glands. Muscles, skeleton, and other tissues soon were differentiated. Finally a skin was formed. Instructions for skin formation included the folding of this outer tissue into the mouth. Because of the infolding, structures of common origin develop: pointed scales on the outside, teeth on the inside. The scales of sharks are of the same basic design as teeth. The infolding is important because the embryonic tissue layer which forms the digestive tract and related organs is incapable of developing

Shark embryo. In this remarkable photograph of the embryo of a shark notice the large yolk-sac connected to the tiny, unborn fish.

hard structures.

When the genes directed formation of the digestive tract, some cells were destined to grow around the globular mass you see in the photo. The young of many species display a bulbous yolk sac hanging from their stomach region, which gradually disappears as the fish grows and utilizes the yolk.

Chapter V. Asexual Reproduction

A species has the best chance to endure when its most successful members can contribute their genetic material in the process of reproduction. There are, in general, two methods of carrying out the process of reproduction. One is for a cell simply to duplicate its genetic material and then divide, giving each cell an identical heritage. The other is for a cell from each parent to divide without duplicating the genetic material, forming cells with half the usual number of chromosomes, and then for one cell from each parent to join. The new cell will then possess a full set of chromosomes. Such a cell is considered diploid, and its two predecessors, with half the set of chromosomes, are haploid. The asexual mode of reproduction does not allow for the great potential variations seen in sexual reproduction, but it also does not require another individual. Being, in a sense, more simple, a greater number of offspring can be produced in a relatively short period of time. There is no energy spent in bringing the sexes together or in competitive battles for a mate. Most organisms, including the unicellular phytoplankton species, simply divide into two individuals, each of which then grows, divides into a total of four, and so on, in a geometric progression.

Some of the simplest forms of animal life, like the amoeba, will simply split in two in a

> "The great majority
> of plants have two alternating
> life cycles, one sexual and
> the other asexual."

process called fission. The new amoeba is identical to its parent. Since the amoeba is a unicellular animal, its reproductive process is identical to the normal mitotic division of all body cells in the growth of multicellular animals. The cells remain diploid.

Other animals, including some fish and a number of invertebrates, will produce an egg which is not a gamete but a diploid cell. It develops without fertilization by a male, and in fact such a species has no males at all. Since all the individuals of the species are females, all of them producing eggs, this mode of reproduction, called parthenogenesis, has the advantage of producing a large number of new individuals in a relatively short time.

Common to some animal forms, especially jellyfish and sea anemones, and common to a great many plants, is another asexual method of reproduction known as budding. In budding, the same mitotic division of cells occurs as in the fission of amoeba, with the difference that the cell divides itself into unequal parts. While both cells are diploid, the larger part is considered the parent, the smaller the bud. Sometimes the bud will grow from the cellular wall of the parent, eventually breaking away to mature on its own. Sometimes the parent will form a protective shell or pouch in which several buds grow and begin to mature before breaking away. In any case, the buds begin as single cells, and they are diploid and require no fertilization.

The great majority of plants have two alternating life cycles, one sexual and the other asexual. The asexual generation, having diploid cells, is usually the dominant life cycle and is called the sporophyte. Sporophytes produce spores. They produce them by the thousands. These unicellular diploid elements are capable of developing directly into adults, and again the process is asexual.

Chains of Plants

The tiny unicellular plants in the picture above are among the most abundant forms of life. Called diatoms, they flourish in both marine and fresh waters. Observable only through a microscope, they make up more than 99 percent of all the plant life in the sea, constituting the base of the marine food chain. Using photosynthesis to provide energy, they absorb minerals and salts and build delicate shells of hemicellulose impregnated with silica. Thus, they literally live in glass houses. Their homes may take a variety of beautiful shapes—bracelets, links of chain, fine needles, minute pillboxes. Their shapes increase surface area and help them float. Few have any mechanism for motion. Most diatoms reproduce asexually, each cell dividing in half—and retaining half the original shell. The new and old cells must then form new shell halves to cover their exposed sides. Since the new halves are made to fit inside the old halves, the cells become slightly smaller with each generation. Finally, a small cell gets rid of its old shell entirely and grows, unprotected for a brief period, before building a new and larger cell

> "Diatoms literally live
> in glass houses.
> Their homes may take
> a variety of beautiful shapes—
> bracelets, links of chain,
> fine needles, minute pillboxes."

wall. Some species produce gametes through sexual division, and when the two male and female gametes fuse they produce a full cell without a shell known as an auxospore—which also eventually forms a glass house.

▲ B   ▼ A

▲ C

Alternation of Generations

In plants there is a multicelled haploid phase. This means that each chromosome is lacking its pair, and that therefore the cell has half the number of chromosomes. In this haploid phase the gametes are produced. In some of the mosses and algae, this gamete-producing generation dominates the plant's life cycle.

A / Tunicates. These animals release sperm and eggs that produce larvae closely resembling tadpoles with elongated tails. Each tiny creature soon absorbs its tail into its head, losing its chordate shape and assuming the jellyfish form seen at lower left. Then, in its asexual stage, the animal begins to bud.

B / Sea lettuce. The life cycle of the sea lettuce or green algae is more complex. The sexual and asexual individuals of sea lettuce appear to be the same. But there are differences here that are not apparent.

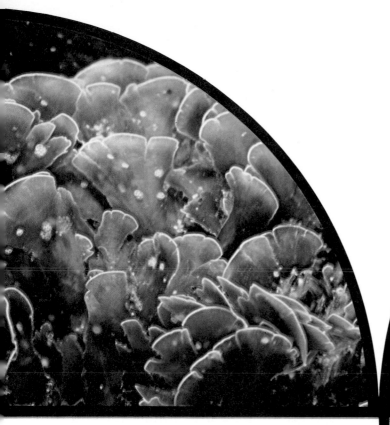

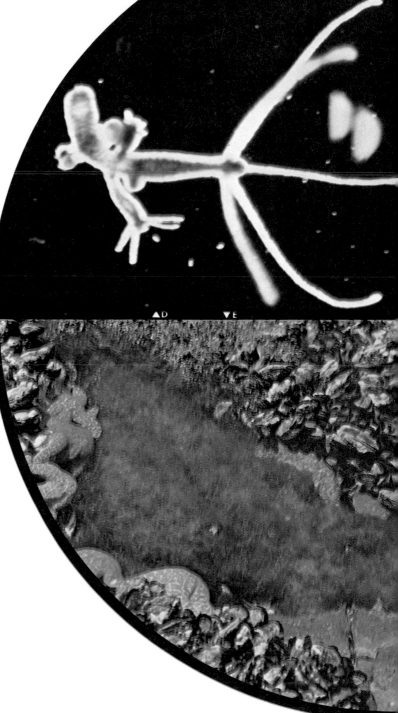

▲D   ▼E

Not only do its adults produce the customary diploid spores and haploid gametes. Some also produce haploid spores. These do not unite sexually with another haploid cell but simply grow by mitosis as would a diploid spore. In such cases, the entire plant remains haploid.

C | Brown algae. These also alternate generations. Their sexual and asexual phases contrast even more markedly in appearance, the maximum growth of the sexual plant rarely attaining more than microscopic size. In the past, the two generations were believed to be unrelated species.

D | Hydroid buds. The pods contain new individuals which are soon released as miniature medusae. These jellyfish rapidly grow in size. Ranging far and wide, the adults, in the animal's sexual stage, release sperm and eggs into the water. The gametes unite and the egg fixes itself on a bottom rock or other surface, there to produce a new vegetative individual.

E | Tide pool. The tide pool is host to many forms of algae, hydroids, and tunicates, as well as a great variety of other plants and animals. It is one of the most productive areas of the sea.

Life Cycle of the Obelia

Here is an unusual animal. Called the obelia, it is one of the hydroids, an animal usually comprising two kinds of polyps, some specialized for feeding and some for reproduction. The animal usually lodges among rocks, and creeping runners may attach to floating wood, mussels, or seaweed.

The feeding polyps are the long ones, thin tentacles that catch minute worms and crustaceans, eggs, and larvae, and occasionally a tiny fish. They nourish the entire colony, from whose base grow the reproductive polyps, stubby and shaped rather like urns. Inside the urns little buds develop, and these are eventually released into the water. The process is asexual, the buds being equipped with the full number of chromosomes re-

> "Bell-shaped, with a mouth for feeding and many long tentacles armed with stinging cells, the animal was named after the Gorgon Medusa, the maiden of Greek mythology whose hair turned into a nest of snakes and who petrified anyone who looked at her."

quired in each cell, and developing by division. This is not uncommon in the larger group of coelenterate animals to which the hydroid family belongs.

But this is only one stage of the spectacular life cycle of the obelia. The little buds are jellyfish. Each of them rapidly grows to its full-fledged size known as the medusa. Bell-shaped, with a mouth for feeding and many long tentacles armed with stinging cells, the animal was named after the Gorgon Medusa, the maiden of Greek mythology whose hair turned into a nest of snakes and who petrified

anyone who looked at her. The name is apt, for the medusa's tentacles paralyze its prey most effectively. Even a human swimmer who doesn't watch where he's going may receive a small shock from its sting. There are both male and female medusae, releasing both sperm and ova in the water. When the egg and sperm unite, and fertilization occurs, a larva forms, and it swims off to the bottom, fastening upon a rock or piece of kelp. What develops? A new colony of polyps! Then the life cycle repeats itself in what biologists call alternating generations.

This lowly hydroid provides an insight into the evolution of its relatives—the true jellyfish and the corals. Most scientists concur that the hydroids are more primitive than their two relatives and sometime in the distant past gave rise to them. The anthozoans (corals, anemones, and sea fans) surely originated from a hydroidlike animal which possessed a reduced free-swimming medusa stage. Rather than form sexual buds to release gametes, the anemones found an advantage in producing eggs and sperm themselves. Adequate nourishment was provided by the stationary adult, while the freely released eggs and sperm provided the necessary dispersal. They and the corals and sea fans remain attached to the bottom.

An opposite evolutionary path was taken by the scyphozoa (jellyfish). They found a free-swimming life-style the best method of feeding and reproducing.

Fragile but menacing, obelias are colonies ranging from one to eight inches in length which grow on seaweeds, rocks, and piles. The medusa, only one stage in the obelia's life cycle, has anywhere from eight to 24 tentacles, and swimmers are warned that they, too, are subject to the stings.

FEEDING POLYP

FREE-SWIMMING MEDUSA

BUD

FULL-GROWN MEDUSA

REPRODUCTIVE POLYP

EGG

SPERM

DEVELOPING MEDUSA BUD

ZYGOTE

BLASTULA

CILIATED LARVA

YOUNG POLYPS

LARVA
METAMORPHOSING
INTO POLYP

Kelp fronds. Globular gas bladders act as balloons to keep the plants suspended in the water and growing toward the surface.

Kelp forest. Due to their rich marine populations, California kelp forests are favorite places for divers. Divers must be wary of the kelp's entangling fronds.

World Champion Growers

Clinging tenaciously to outcroppings of rock, this massive streamer grows more rapidly than any other plant on earth. It is *Macrocystis pyrifera,* the giant kelp of California. A form of brown alga, it can spring up from the ocean floor at the rate of over a foot a day, its great stalk supported in its climb by tiny round bladders filled with gas that form at the base of saw-toothed fronds. The plants may grow as long as 200 feet.

The giant kelp is one of a great many plants with alternating sexual and asexual generations of a kind similar to the cycle we have seen in such animals as obelia. Most of its astonishing growth is accomplished in a diploid stage (all cells containing a full set of chromosomes), in which the plant reproduces asexually through division, releasing diploid cells growing into new adults. At the same time it will produce haploid cells (half the full set of chromosomes). These cells

> "Each plant may grow to a length of 200 feet in five or six months, and die, to be immediately replaced by new stands."

grow without fertilization into a microscopic haploid plant which then releases the appropriate male or female gametes to form the common giant kelp.

Parthenogenesis

A different reproductive system is found in the *Poecilia mexicana*. There are two kinds of females in this species, one producing the usual mixture of male and female offspring, the other giving birth only to females. Some of the male's characteristics are transmitted to the unisexed infants.

Parthenogenesis is the general name for this mode of reproduction. The Amazon molly represents an extreme case, a species in which there are only ladies. None of the individuals is male. In this species there are eggs, but they contain the full complement of chromo-somes. The eggs are formed by simple division. They require no fertilization by a male. They develop nicely without it. Naturally all the eggs develop into females, all of which in turn produce more eggs. The advantage: a large number of individuals are produced

> "There are two kinds of females in some species, one producing the usual mixture of male and female offspring, the other giving birth only to females."

continuously. If there were males in the species, so only half the population produced eggs, they might not survive.

An interesting aspect of fertilization is that the male spermatozoan accomplishes two different things. On the one hand, it initiates development of the egg. Very few species can have their eggs fertilized without sperm of some sort. And the Amazon molly's egg will not develop without the attachment of a spermatozoan to provide the stimulus for cell division. On the other hand, the sperm also contributes the paternal component of genetic character by sharing its chromosomes with the female cell. In pure parthenogenesis, found in some arthoprods and other invertebrates, the individual carries both sperm and egg within itself, and thus stimulates development of the egg by itself. The

Top minnows. *Some fish, like the top minnows above, reproduce by a method in which the eggs develop without male fertilization.*

Amazon molly is not so pure. She has no sperm of her own, but still must consort with males. So—the males are of another species, usually the common molly or sailfin molly. These males contribute their sperm, and the eggs of the Amazon molly are given their initial thrust toward development. But because these eggs have a complete set of chromosomes, there is no fusion with the male nucleus, and the sperm is rejected.

Chapter VI. Sexual Reproduction

Early in the history of life on earth a method of reproduction developed which was in some ways better than the simple asexual method. This was a means of sharing genetic information so that valuable traits could be spread over a larger population and new genetic combinations might be formed. In these sexual reproductive methods, cells from different individuals combined to form nuclei unlike any others before them. But sexual reproduction creates its own problem: two cells from different individuals, a male sperm and a female egg, must come into contact for fertilization. Many creatures, either by physical forms of communication or by instincts we do not yet fully understand, come together at an exact time and place for the simultaneous shedding of sexual products. Close proximity of great numbers of eggs and sperm guarantees successful fertilization. The male of one species may first select a fixed location and then force the female to it, not allowing her to leave until she has laid her eggs, which he then fertilizes and guards. Other males take no chance at all, but graft themselves permanently to the female body. Some species are hermaphroditic, carrying both sperm and eggs simultaneously. While they usually cross-fertilize one another, they often do not. Many of the more advanced creatures carry out internal fertilization, which provides the greatest chance for egg and sperm union as well as protection of the embryo.

Along with the development of more sophisticated methods of sexual reproduction have come greater measures of parental care, providing the young with better nourishment and greater protection over long periods of time, and in the case of some aquatic mammals making use of a high intelligence to teach the young how to swim, how to breathe, how to hunt, how to find food.

When two squids mate, the male uses one arm to grasp his sperm packets and introduce them into the female's oviduct, thus performing internal fertilization. In the same family as the octopus (both animals are

> "Many creatures, either by physical means of communication or by instincts we do not yet fully understand, come together at an exact time and place for the simultaneous shedding of sexual products."

molluscs along with, among others, clams and snails—which shows how diverse nature can be in the area of variations, given a couple of hundred million years), the female squid is not as conscientious in the performance of maternal duties as is her eight-armed cousin. She lays her small eggs on the bottom in masses of jelly shaped like cigars; they are called "dead man's fingers." The giant of this species, the deep-sea squid, is the largest known invertebrate—with a body as long as 18 feet and arms as long as 30 feet. Giant squid come even bigger than this, for we have found parts of larger animals in the stomachs of sperm whales. These are the monsters who leave scars all over the heads and bodies of older sperm whales, souvenirs of titanic undersea bouts we would give much to witness.

Squid. Squids are muscular, jet-propelled molluscs which have sexual reproduction. Here, three males are attempting to mate with a female. After fertilization the females will lay great gelatinous masses of eggs on the bottom.

Fertilization

The open sea, with its unpredictable shifts of current and the constant threat of hungry predators, makes the task of development of an egg far more hazardous than it would be if the eggs were secured inside their mother's body and the sperm delivered there. With the exception of dolphins, whales, and other mammals, and of some lower animals like the octopuses, females of marine species are not equipped to receive sperm and fertilize their eggs internally. Sharks, certain rays, and sea snakes have in-

> "The open sea, with its unpredictable shifts of current and the constant threat of hungry predators, makes the task of development of an egg far more hazardous than it would be if the eggs were secured inside the mother's body and the sperm delivered there."

tromittent organs which transmit sperm directly to the female and make internal fertilization possible. Male mosquitofish have an anal fin modified into a gonopodium that performs the same task. But the great majority of marine animals simply shed their eggs and sperm into the ocean. Even species that have pouches in which their eggs develop, such as the seahorse and pipefish, must fertilize the eggs externally.

Fertilization is the critical event in sexual reproduction. On page 46 we have seen that the sperm performs two functions, contributing genetic material to the embryo and providing the stimulus which commences cell division and the growth of the embryo. The two processes are not the same. Egg development can be stimulated, either arti-

ficially or by foreign sperm, without the contribution of genetic material. The processes are also irreversible. Once the embryo has received the stimulus to begin growth, development cannot be stopped. If this process has begun without the presence of new genetic material, the material cannot later be added, and in such rare cases subsequent development is abnormal.

Most spermatozoa are equipped with a head that contains the genetic material to be added to the female cell, a middle section with mitochondria that control their own metabolism, and a flagellum, or tail, which gives them mobility. After spawning, the eggs of most species float in the water, immobile, or rest on the bottom. They usually have a gelatinous outer coating which dissolves in the water. As the spermatozoan comes into contact with an egg it must penetrate the egg membrane. Its head contains a granule of enzymes. On contact, this granule forms a filament attaching the sperm to the egg. The spermatozoan next slowly enters the egg, which responds sometimes by changing color, sometimes by forming a protruding fertilization cone with its cytoplasm, enveloping the sperm and drawing it further forward. After the sperm has completed entry, the egg's vitelline membrane expands, forming a tough new kind of membrane.

Sea snakes, reptiles whose ancestors returned to the sea, have adapted well to their new habitat. Most of them need not return to land to lay their eggs, unlike their other reptilian relatives the turtle and Galápagos marine iguana. The sea snakes stay in the sea to copulate, the eggs develop within the mother, the young hatch, and the female bears live young. Sea snakes are poisonous to both fish and warm-blooded animals, but fortunately seldom attack man. Most have heavy bodies, tiny heads, and broad tails.

Egg release. *This serpulid worm is releasing eggs. These will receive no parental protection, relying on chance for fertilization and successful development.*

Chemical communication. *These feather-duster worms cannot move and must communicate chemically to achieve simultaneous sperm and egg release.*

External Fertilization

Serpulid worms live off Catalina Island. They are sometimes called feather-duster or spiral worms and belong to the group of segmented animals called annelid worms. Unlike the common earthworm, they are polychaetes, a group comprised mostly of marine worms, often brightly colored and with bristly appendages.

Serpulids live below the low-tide mark along most tropical and temperate shores, growing to lengths of three or four inches, fastening itself to a shell or embedding itself in coral or rock. It builds a hard tube, irregularly coiled, from secretions of limestone, and in this tube it lives excepting at mo-

ments like this when it reproduces. From the end of its tube it protrudes feathery gills, often beautifully colored, trapping particles of food that fall from the surface plankton. If danger approaches it simply withdraws its plumes, completely closing the entrance to its tube as effectively as if it had a stopper.

Serpulid reproduction is sexual, by means of sperm and eggs, and because these worms are not highly mobile their union takes place pretty much by chance. Here the wave surge is seen to carry the eggs off in first one direction, then another. They must rely on external fertilization to bring the gametes together to commence the process of growth. Synchronous release of eggs and sperm is essential to ensure fertilization.

Taking No Risks

The horseshoe crab (above), an ancient animal of the North American east coast—harmless, but for us monstrous in appearance—runs few fertilization risks. The mature male is equipped with two strong hooks at the sides of his mouth, possessing an exclusively sexual function. In springtime the female of the species begins to move from deeper waters toward the shallows. By means of his hooks the male grabs the female by the hind part of her shell and hitches a ride. More than one male can join the female. Another male can clamp onto the first and so on—until a chain of as many as four or five has formed, holding on for days or weeks. The female drags her escort onto a beach at high tide, digs a hole for her eggs—and her males fertilize them. The horseshoe crab has remained unchanged for at least 175 million years. He is not a real crab at all; his closest living relations are the scorpion and the spider.

Part of the Mating Game

The grunt sculpin would rather skip along the bottom than swim. They accomplish this by use of projections of their pectoral fins which allow them to spring forward in a series of hops. When required to swim, they move like balloons. Add to this the fact that grunt sculpins have large heads and eyes that move independently and a man might be entitled to view them as animal comics. The mating ritual takes place during the months of August through October. It begins with a female chasing a coy male about until she forces him into a crevice or other chamber. Once inside, the female blocks the entrance. She wants to be sure the male will be available to fertilize the 150 or so eggs she will deposit on the chamber walls. After he does so, the female allows him to leave. How long the eggs incubate depends on the temperature of the water. They live in the icy waters of the Bering Sea and all along the Pacific coast from Canada to California.

The Mystery of Instinct

Wrasses off a reef near Baja California. Hundreds of them. There are thousands more mating along other reefs. Is it a certain point reached in the lengthening of the daylight hours? Or a particular combination of temperature and light occurring periodically during each year? By what biological rhythms or subtle clues from the environment do millions of these and other fish achieve the proper development of sperm and eggs and simultaneous spawning? Like clockwork, all at a time, they aggregate and reproduce as surely as the sun rises and sets.

We have not yet unraveled all the mysteries of instinct. But we do know that the wrasses' reckoning is exact, their timing precise. In the vast open reaches of this ocean, it is no simple task to fertilize an egg. A whale may swim about for many months without meeting a female. The anglerfishes inhabiting the

middle depths are so rare that they could only perpetuate their species by the male's becoming a parasite, permanently attached to the female. Only thus will he be on hand when her eggs are ready for fertilization. Sometimes animals communicate with each other visually, sometimes by making sounds, and sometimes by releasing in the water chemicals that signal members of the opposite sex to join in the act of procreation. But if eggs must be externally fertilized, the best way to maximize successful encounters of

Wrasses. There are about 450 species of wrasses living in the waters of the world, most of which change color at different stages of their development. Most of these varieties have prominent canine teeth and are carnivorous, frequently voracious. They range in size from tiny cleanerfish to over a few feet in length and 200 pounds in weight!

egg and sperm is to shed them all at the same time and in the same place. By whatever fine and wonderful mechanism they have developed, this is just what these wrasses have been given the capacity to do.

Internal Fertilization

Stingrays have eggs which develop and hatch within the female. Their embryos do not take nourishment from the mother's bloodstream, as the young of man and other mammals do, but from the yolk sac of the egg. Until an egg is ready for hatching, it is kept within the mother in what amounts to a brooding pouch. This gives it considerably more protection than one released in the environment. Since no eggs are lost, fewer of them are produced—and because the eggs are so large often only one is produced at a time by each female.

To accomplish internal fertilization, the male ray is equipped with small clasping pelvic fins which grow rigid and come together like hands at prayer to form a channel down which the sperm runs into the female's ovarian cavity. Rays, like their primitive cousins the sharks, have remained relatively unchanged for millions of years. During the past 100 million years they have shared with sharks the same highly advanced set of reproductive techniques. In evolutionary terms, they have in this way successfully competed with the thousands of "higher" fishes that followed them and surpassed them in other areas of development but still shed their sperm and eggs into the sea, where they are at the mercy of the environment.

The Mating Dolphin

It is early morning, and these two dolphins are making love. The male will nuzzle and bite the female, rub his body against hers, and caress her with his flippers and dorsal fins. She will respond in the same manner. Then she will dart away; he will give chase. They will tear through the water at speeds approaching 35 miles an hour, rushing upwards to break the surface in soaring arcs, splashing back down, knifing through the depths in different directions, pulling together again. Soon she will grow quiet. She will turn on her side, he on his. And swim-

> "Eleven or 12 months will pass before she gives birth. Chances for her to have twins are about the same as those in a human birth."

ming along quite gently together, with no violent thrust of body or tail, they will mate.

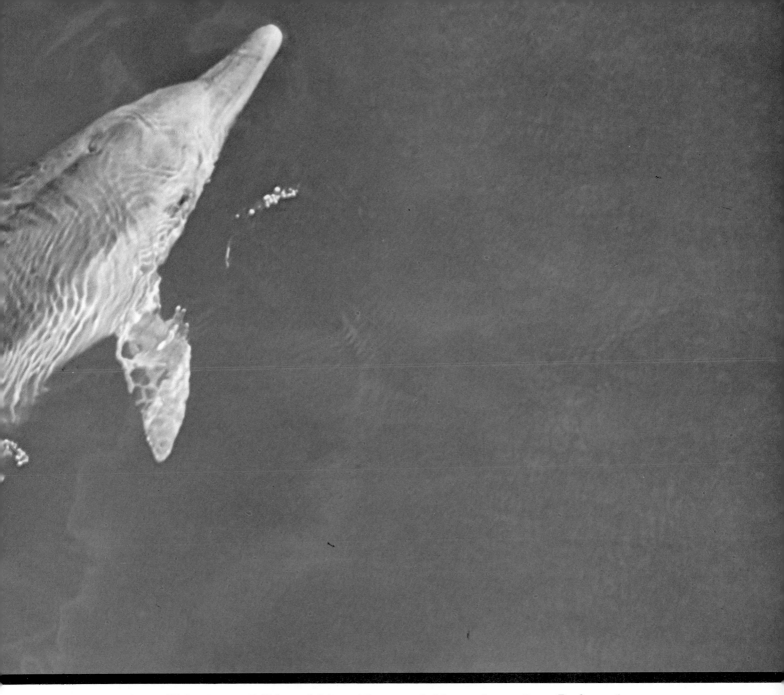

The mother will bear one child, and 11 or 12 months will pass before she gives birth. Chances for her to have twins are about the same as those in a human birth.

Born tail first, the pup must immediately learn to breathe, for breathing in dolphins is not involuntary as it is for us. The mother, or occasionally another female in the pack she has joined, takes the newborn child to the surface immediately. The instinct is so ingrained that whenever a dolphin is sick, or calls for help, another dolphin will nudge it quickly to the surface. In fact, these community activists may perform that duty for other animals unrelated to them. We heard of the case of a female dolphin that became concerned over the dead carcass of a large tiger shark, and, pausing only infrequently to eat, kept nudging it to the surface for eight full days, until it began to decompose! If such deviations of instinct are possible, then there may be some truth to the stories we hear of men and women saved from drowning and pushed to shore by friendly dolphins.

Hermaphroditism

Hermaphroditism is the condition in which an individual animal possesses the reproductive organs of both sexes. Although some hermaphroditic species are capable of self-fertilization, it is more common that individuals cross-fertilize one another. Hermaphroditism can increase the reproductive potential of a species, since all the individuals produce eggs instead of only half—as is the case in species with separate sexes. Also, in solitary or rare species it guarantees that whenever two individuals meet mating can take place, because each individual has both sperm and eggs.

Hermaphroditism is the normal mode of reproduction in many animals. For instance, the sea hare is a marine snail without a shell. At mating time one sea hare climbs on the back of another. Both face the same direction with the one behind fertilizing the eggs of the one in front. A third may mount the second so that its eggs are being fertilized while it is supplying sperm to the sea hare in front. In this manner long chains of mating individuals may form. Sometimes the first individual will swing around and mount the last so that the circle is closed. All the mating animals are thus simultaneously fertilizing eggs, plus being fertilized themselves. Some species don't rely on locating others of their species. They are capable of using their sperm to fertilize their own eggs. There are certain drawbacks but self-fertilization assures continuation of these species.

Nudibranchs. On the page opposite, we see two nudibranchs mating. Each animal has both eggs and sperm, so each will bear young after exchanging sperm. Hermaphroditism is common in gastropods.

Hermaphroditism and Sex Change

Please see pages 64 and 65 for these fish.

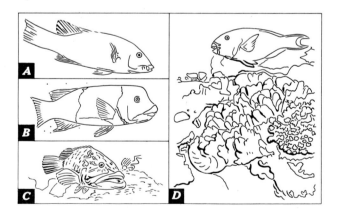

A / Female sheepshead. All large, brightly colored male sheepshead have gone through a female stage in which they produce eggs.

B / Male sheepshead. This fish was formerly a female. In the sheepshead there are two types of males. Most of the males are small, similar in color to the females. However, some males are much larger and have a different, often brighter, color pattern. Examination of the sex organs shows that the small males have normal testes, and have been males since birth. However, the large, differently colored males have degenerating ovarian tissue in their testes. They are females which have transformed into males.

C / Sea basses. In many sea basses the individuals are both male and female, and self-fertilization may occur. However, groupers and some other sea basses exhibit another type of hermaphroditism. All groupers begin life as females and, after functioning as females, transform into males. So—all large groupers are males.

D / Parrotfish. Parrotfish and wrasses in a sense have three sexes: females, males, and males that are sex-transformed females. Both types of males are capable of fertilizing eggs. In one species when the male parrotfish is separated from his harem, the dominant female of the group changes sex.

▲ A   ▼ B

▼ C   D ▶

Please see page 62 for descriptions of these fish.

Chapter VII. Peaceful Romance

Fish and other sea creatures in search of mates are equipped with many methods of attracting members of the opposite sex. Some marine creatures have the ability to change color as they seek partners. Humans, not having such natural ability, clothe themselves in garments to catch and please the vagrant eye of the coveted girl or boy.

Love dances—the grace, beauty, symmetry of which can bring to mind a Viennese waltz—are practiced by many species. The male dolphin performs in any of dozens of ways for the female of his choice—often contacting and caressing her body to stimulate her interest, or displaying his skill and agility as a swimmer by swimming directly at her, artfully swerving at the last possible moment to avoid collision. While engaged in this behavior, they are very vocal.

The sense of smell also comes into play during the mating ritual and serves an important function in bringing the sexes of certain species together. Chemicals, called pheromones, are given off by some females and broadcast in the water by currents. Males of the species pick up the odor. Their ardor aroused, they seek out the female. The chemicals given off must be of small enough quantities to be carried readily yet of sufficient strength and singularity to be immediately identifiable by the males. In case of nonmobile creatures which cannot get together to mate, the *only* means of sexual communication is by the discreet and timely use of chemicals.

The ability of certain sea animals to phosphoresce is another method used to attract mates. When seeking a lover, these fireflies of the sea flash their lights at certain frequencies or in specific patterns. The result is clear: members of the same species are informed of the desire to mate, members of other species remain in ignorance.

Many animals in the sea are capable of changing color. In the pursuit of their normal lives, this ability gives them a better chance to escape a larger animal or, conversely, a better chance to capture an un-

> "Many fish grow loquacious during the mating season. Whales probably serenade their chosen mates with clicks, grunts, and low moans— music to harkening members of the opposite sex."

wary prey. But in mating season the males of many species blossom forth in spectacular arrays of wedding displays. Normally drab creatures take on hues of brilliant intensity, making of themselves conspicuous advertisements. This gaudy appearance attracts females ready to mate. It also alerts similarly colored males, and territorial fights occur between these sexually activated creatures. Tests made on the stickleback show that a crudely made model in mating colors will be attacked more readily than an accurate likeness in everyday dress. The female stickleback, ready to mate, turns a silvery color and bulges with eggs. Again the male will choose a crudely made model with a bulging profile over a precise, thin one. When the excitement of mating has passed, the colors fade.

Cuckoo wrasse. The cuckoo wrasse is capable of changing color for many reasons, among them camouflage, day-night cover, and sexual differentiation. The pictures on the opposite page show the animal's mating colors during sexual display (top), and normal, out-of-breeding-season colors (bottom).

Octopus Birth

This photograph of two octopus undergoing their mating ritual marked the beginning of a long series of studies conducted by our divers and scientists of the Oceanographic Museum in Monaco. Fortunately one day as our divers were making their routine environmental surveys, they discovered this mating in the waters near the museum. After the female rejected her mate, we were able to locate a small cave where she had established a home. It was in this nursery that she remained alone, waiting the five or six weeks for the hatch. Periodically we sent divers down to visit. They found that she had glued her thousands of eggs to the ceiling of the nest—50 clusters containing up to 4000 embryos each. The mother-to-be shielded them with her outturned arms, kept them oxygenated by squirts of water from her funnel, and vacuum-cleaned them with her suction cups. We offered her a fish but it was promptly turned down; the female octopus stops eating at brooding time. Soon she was near starvation. It is unusual for an octopus to leave her eggs, but she attacked our men. During her pursuit of one of them, she gave us an unexpected opportunity to remove a strand of her eggs for study. Having cleared the intruder away, the octopus returned to her nest.

Back at our Center of Advanced Marine Studies in Marseilles we used special camera equipment to photograph the miracle of octopus birth. We could see the great eyes and pulsating mantles of the infants and the babies' struggles to break their bonds and begin life on their own. Finally it happened. The babies burst from their egg cases— translucent forms duplicating that of their mother. Down in the ocean the same feat was taking place. The mother was dying, but blowing her young out of the nest, releasing them to the world. Few survived the first hours. In increasing numbers fishes had gathered outside waiting to gobble up the newborn. Of the octopus's 200,000 or so offspring only one or two would reach maturity and reproduce in their turn.

The Dancing Cod

"Peaceful romance" doesn't always imply passivity. The male codfish is a dancer. First he establishes a territory. In itself this is not all that peaceable a job. But once the territorial business has been settled the male cod becomes an active lover. Patrolling his ground vigorously, making sure that other males and unripe females give it a wide berth, he waits for a female who is ready for him—a lady of his species that moves into his water space with serene aplomb. Now the male goes into his act, raising his back fins, after a moment or two closing in on the female. Perhaps ten inches away from her he erects his median fins and contorts himself. The body undulates, the fins flap: this is the courtship display. If he has picked the right partner, she begins to "listen" to his enticements: his noises, dancing, displaying.

His point: to get her to swim to the surface where cod mate. At last, after much maneu-

> "Patrolling his ground vigorously, making sure that other males and unripe females give it a wide berth, he waits for a female who is ready for him— a lady of his species that moves into his water space with serene aplomb."

vering and "dancing," he achieves his goal. Near the surface the male mounts the female. As described by V. M. Brawn, "he immediately slipped down one side of the female, still with his ventral [belly] surface closely pressed against her body and clasping her with the pelvic fins. The male came to be in an inverted position below the female and with the ventral surfaces of both fish and their genital apertures closely pressed together." The female now spawns, her eggs are fertilized, and she swims out of her mate's territory.

By this courting behavior, the fluttering of fins and sinuous movements of his body, the male has lured his partner away from the near-freezing waters where adult cods spend most of their lives. Her eggs would stand no chance of hatching in the cold, hostile environment of the bottom. She has spawned at the warmer surface, and her lighter-than-water eggs can hatch there in about ten days. The little fish will stay in the rich planktonic layer, feeding and being fed upon for about two and a half months. The survivors will then sink to the bottom where they will live and grow to maturity.

All this sounds peaceful enough, I suppose, but the male cod's defense of his right to breed has its forceful side. Large cod "display" to rival males in many ways during the course of establishing territories. Like other animals, they specialize in faking bigness: by expanding their gill covers, puffing up their mouths and throats. They grunt; they butt; they pretend to bite. In any single volume of water space the smaller flee, the bigger sort themselves out into a pecking order much like that of a barnyard. When you think about it, you see that this belligerence—very little of which, in any event, ends in damage or death to the individuals involved—has its survival aspect. Chased from familiar territory by superior males, the inferior male must seek elsewhere for food and love. Often his wanderings will take him far afield, opening up new regions for his species—promoting its survival.

The ritual. *The male cod performs, gains the interest of a female, and gets her to swim with him to the surface where they can mate.*

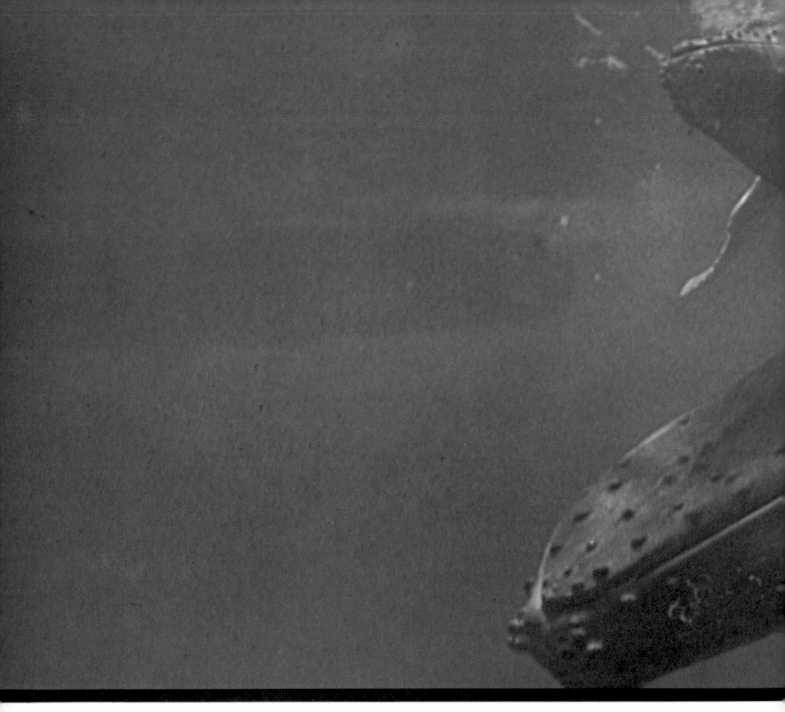

The Singing Whale

In the days of wooden sailing ships seamen were often treated to song recitals of humpback whales amplified through the hulls of their vessels. Then steam and diesel engines drowned out the whale's song until only a few yachtsmen knew of it. Tragically, just as the rest of us are beginning to record this music, the singers are growing rare.

The humpback is a filter-feeding whale which migrates to the Arctic and Antarctic oceans when summer reaches these regions. It inhabits all the world's oceans. We have studied the animals near Bermuda where they pause in their travels—we think to form pairs, if not to mate. Their behavior in these warm waters is playful—they slam their tails on the water's surface, wave and slap fins, jump out of the sea in spectacular leaps.

Their songs are a long plaintive melody, stretching from high and low notes, repeated in sequences that vary widely with the

singer. They may be seven minutes long, or 30, and sometimes different songs are worked into performances lasting for hours. The whale may repeat the same song over and over with minor variations. There is no set pattern for humpbacks; still, spectrographic analysis has shown their songs are basically similar.

Why do whales sing? We aren't sure. We don't know if the sounds are produced by male or female, whether we are hearing a chorus, duet, or solo. They are loud, and

Humpback whales. These mammals, averaging 48 feet in length, are easily recognized by the roundish lumps on their heads and their long, whitish-blue pectoral fins—analogous to man's arms. Like so many other whales, unfortunately, they are in danger of extinction.

sound travels far underwater. We feel sure the songs are meant to announce the whale's presence to others of its species. At the proper season they are sung to invite companions or to propose to a mate.

Messenger Service

Pheromones are gentle fragrances precisely conveyed to species-neighbors and potential mates. The message may dispel fear of territorial threat or, on the other hand, produce it. In some species the chemical includes information on what species the intruder is, his social status, his sex, his age or size, his reproductive state or his intentions. This is all vital information.

The chemical form of communication is especially helpful to creatures living in the unlighted deeps of the abyss and the murky coastal areas where vision is restricted or nonexistent. And consider its importance to the bottom-attached animals who, by the nature of their design, are unable to participate in mating rituals. Without some form of communication, mating and reproduction would be haphazard at best. Many species could not have evolved without such

> "Pheromones are gentle fragrances precisely conveyed to species-neighbors and potential mates. It is believed that the chemical includes information on what species the intruder is, his social status, his sex, his age or size, his reproductive state."

communication. The oyster, for example, has an amazing method of reproducing based on sensitivity to chemicals. The male oyster can release its sperm on the call of many stimuli—among them the sperm of unrelated invertebrates. The female oyster doesn't release her eggs until she detects the presence of oyster sperm in the surrounding waters. The sperm-laden water is taken in and fertilizes the eggs inside her shell.

In certain species of crabs and crayfish the pheromone released by females is the same hormone that causes molting. As molting time approaches and the hormones are released, males are attracted to the premolt female. The aroused male will seize the female, holding and defending her against competitors and predators. When she has lost her shell, mating occurs. The male will continue to protect the now pregnant and extremely vulnerable female until her new shell begins to form, helping the species in two ways: the living female is not lost, the fertilized eggs she carries are sheltered. It is the response of the male to the premolt pheromone release which ensures that he will be on hand when molting occurs. Since the female is still in her shell when males first arrive, she is able to reject suitors she feels are inferior as well as fight off predators which might be in the vicinity. It has been shown in experiments that some species are able to remember for months the scent of a specific intruder—some catfish react to water from the intruder's tank as though it were there, presenting a threat as it had done once before.

Because the receptors for recognizing pheromones are extremely sensitive, man's pollution of the sea poses an enormous threat to animals depending on chemical communication in their own life processes.

Man may find uses for the pheromones—in marine farming, for example. They may be used to repel predators, to attract males with artificial baits for fishing, to inhibit aggression and cannibalism, and as growth stimulators.

Spiny lobster. This crustacean lives in the Pacific and tropical waters and is more colorful than its clawed relative of northern Atlantic waters. The color does not assist the lobster in its mating ritual, however. It is believed that pheromones help these animals locate and identify mates.

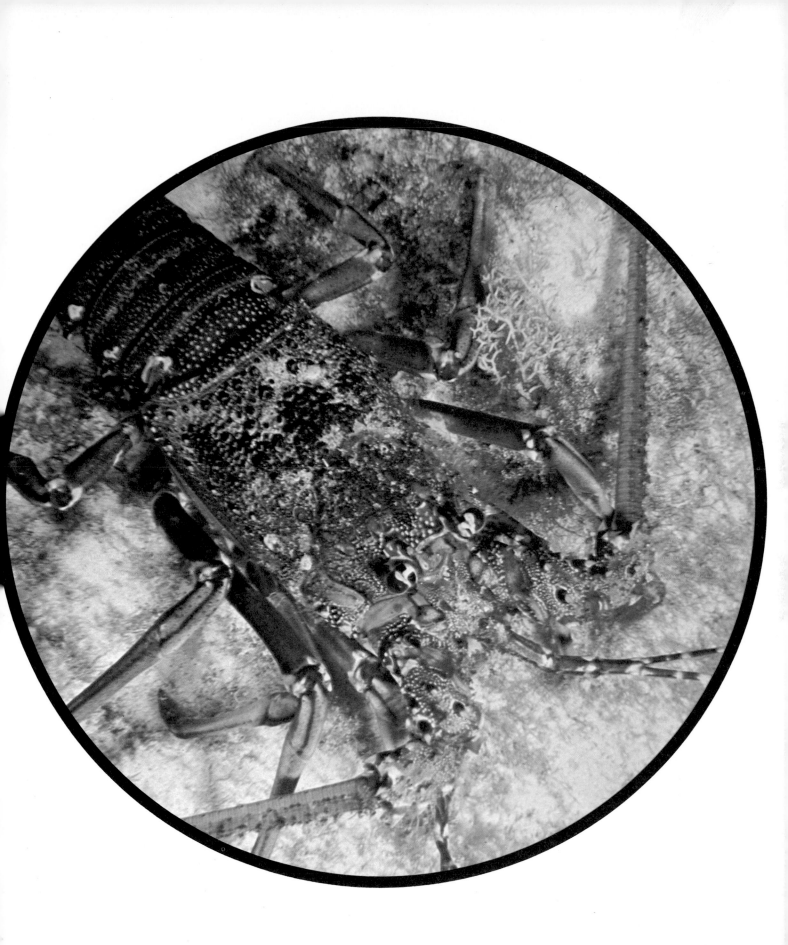

Love in the Abyss

Deep-sea animals have evolved special systems to survive in their inhospitable domain: mouths that can gulp prey larger than the predator, eyes with their own light, or lighted baits. They have also adjusted their reproductive habits to fit conditions in the deep. Some of the brotulids bear live young, while the North Pacific halibut lays its numerous eggs at a depth of about 900 feet in the path of a counterclockwise eddy. This current carries the eggs to shallow water nurseries where the little fish spend their first years before moving to deep water.

Finding a mate in the abyss, where no sunlight reaches, has been solved in a number of ways. The rat tail, a relative of the common cod, vibrates its swim bladder to pro-

> **"The majority of deep-sea fish have the ability to produce light directly or indirectly."**

duce grunts, booms, and other sounds. This fish has a very sensitive lateral line which alerts it to nearby movement.

Although there is no visible sunlight that penetrates below 2000 or 3000 feet, there is other light in the deep, produced by the animals living there. Vision is still important to most deep-sea fish. In a few cases, the eyes have atrophied as the eyes of cavefish have done, but generally they have grown hypertrophied. A majority of deep-sea fish have the ability to produce light directly or indirectly. Bioluminescence is often produced by bacteria living symbiotically on the host. Other fish are able to make their own light with special glandular cells called photophores. The patterns of lights vary with the species. Some of the animals are able to turn the lights off and on to flash a variety of colors: white, green, orange, and red. In some species only the males have lights, leading us to believe they are as important as color changes are to shallow-water fish.

One of the most amazing and practical forms of deep-sea mating is performed by the anglerfish. The eggs of this fish are buoyant, and an attached gelatinous envelope makes the larva appear larger than most of the other members of the planktonic community where they are hatched. The young male has efficient tubular eyes and an enlarged olfactory organ. His sexual development advances rapidly and he begins seeking a mate while he is still living at the surface. He follows the scent of a female anglerfish avidly, for he has a lot of competition from other males of his species, and unpaired he is incapable of making a living in the abyss.

When he has found his female, a fish that can be as much as 25 times his size, he grips her body with specialized small teeth. His position on her body seems to be of little consequence. Once attached, the male changes in remarkable ways. His lips and mouth tissues fuse with the female's tissue, and his alimentary tract degenerates. In time their bloodstreams intermingle and he loses his now useless eyesight. Two small openings remain where the mouth was to allow water to enter for respiration. The male has become a parasite, but he apparently puts no strain upon the female, for some have been known to support as many as three males.

Female anglerfish with parasitic males. In this drawing, two small male anglerfish are gripping the body of a much-larger female anglerfish with their teeth. After a while their bloodstreams intermingle. The male has become a harmless parasite upon the female.

Chapter VIII. Violence for Sex

In the deceptively calm sea most species fight not only for the right to live but also for the right to mate. In the sea world, the fittest and luckiest may reproduce.

In spite of the ferocity of battles for mates or territory, few of them end with one of the participants seriously injured. Fewer still result in the death of the loser. In the sea, while the weak or less fit may be deprived of mating privileges, they are not systematically deprived of life.

There are several reasons why animals fight. First, to seek a "territory," to establish its

> "In the sea, while the weak or less fit may be deprived of mating privileges, they are not systematically deprived of life."

limits, to defend it against all who would take it away. Territory, for nest builders, is essential to success in reproduction. The choicest possible area must be found in order to attract a mate and to provide for the young. Nest-building fish, regardless of size, defend their territory with great boldness. Jawfish, for example, erect their fins, open their mouths incredibly wide (making them look larger to an intruder), and quiver their bodies. Should these methods fail to frighten the challenger into retreating, they may resort to butting, tail slapping, and to biting to ward off the assault on their territory and consequently on their right to mate.

Looking like knights of old, armor-clad crabs also battle over territory. At the approach of an aggressor they elevate themselves on their legs, again to appear larger than they are, hoping to frighten away the intruder before it is necessary to come to grips with him. If this ploy does not work,

the defender approaches the attacker with raised claw. Should this brandishment still fail, the animals engage each other by clinching claws, the one attempting to overturn the other. At last one retires.

The fierceness of the fight doesn't necessarily relate to the size of the combatants. Tiny tube worms engage in violent territorial battles—when an occupied tube becomes attractive to an interloper who tries to move in. If the rear entrance has been used to gain access, the intruder may first encounter the hind portion of the occupant, which he promptly bites with strong pincers. The defender may turn around while remaining inside the tube, thus exposing himself, undefended, to a flank attack. Or he may prefer to leave the tube and reenter, so that the two worms are face-to-face. They then may thrash violently around in combats lasting from 15 seconds to four minutes. They lock together, gripping each other with pincers, pushing mightily back and forth—until supremacy has been determined. Occasionally the fight ends with both worms occupying the same tube, facing opposite ends.

Cephalopods—with their many arms—tend to engage in ritualistic warfare for the right to mate. Before choosing mates, squids and cuttlefish ceremoniously rush at each other, making no contact. When mates have been chosen, the male will defend the female. But fights are often but not always avoided because the smaller combatant withdraws.

Battles for mates and territory may seem to jeopardize species, since males (and sometimes females) can die or be incapacitated in the course of them. But they are necessary. A harsh law in the sea is that only the most capable or fit should win the opportunity to breed.

Bloody Courtships

From the blood on the coats of these walruses, it appears that they have to fight for sex. But they do not form harems or generally bully their females as do their seal relations. (Perhaps because males and females are closer in size—and because the female possesses tusks almost as impressive as the male's.) There are other distinctions. The female walrus is a most solicitous mother. Often giving birth to her pup in the water, near a spit of land or an ice floe, she carries him around on her back when he wearies. She teaches him to sleep in the water, to scoop up the crustaceans and molluscs he likes to feed on from the sea bottom as far down as 250 feet. She doesn't actually wean him for about 16 months, which means that most female walruses can breed only once every two years. A system of strict sexual segregation governs the walrus breeding ground: adult males huddling and piling in one group, females forming another herd with young and immature animals. (For reasons not understood, body contact between individuals is important to walruses.) The walrus matures sexually at about five.

The battle. The blood on the coat of the walrus is evidence of a battle he has just had with an adversary over territory or a suitable mate.

A Primitive Battle

Far back in the haze of history evolved one of the most enduring creatures of the natural world: the sea elephant. For 20 million years the lives of these specialized animals have remained basically unchanged, a primitive drama in four acts: Combat—Mating—Birth—Death. The main arena for this living theater is a remote Pacific island off Mexico's west coast, Guadalupe, a true oceanic island surrounded by warm waters, the last refuge of the great northern sea elephants to which they came centuries ago after they had been hunted off their traditional mainland beaches.

The sea elephant is a nomad, spending more than seven months of each year in the open sea. He must come to land to mate and breed—and it is here, at the beginning of each breeding season, that violence breaks out. The bulls, some of them three tons or more in weight, throw down noisy challenges to one another. The ensuing combats are unpredictable—and can be earthshaking. With short, massive fighting tusks, bulls slash at the soft, unprotected flanks of their opponents in lunges that would instantly kill a man. Many of the fights seem to be primarily tests of strength and stamina—10,000-pound shoving matches in which the object is to destroy the enemy's will. In the water the wounds inflicted can be more serious—vicious piercings of the animal's chest armor of tough hide. Time and again I have seen combats ending as suddenly as they began. Losers do not have to fight to the death. And in any case the huge mammals, primarily adapted to the buoyancy of the sea, cannot keep it up for more than 15 minutes or so. In less-populated rookeries a big, victorious bull could probably definitively establish himself as "beach master" of a harem for an entire breeding season. Today, in the packed conditions of Guadalupe, winning bulls are constantly being challenged—and even overthrown. I suspect that Guadalupe males fight merely out of the urge for combat.

The birth of one new generation and the procreation of another take place in the midst of the chaos. Most births occur at night. For the cows it marks the end of 350 days of gestation—mostly at sea. Each pup is born into an anarchic melee of fighting

> "With short, massive fighting tusks, bulls slash at the soft, unprotected flanks of their opponents in lunges that would instantly kill a man."

and mating. Nearly blind at birth and too weak to move about, his cries of hunger drowned out by the cacophony of the rookery, he may even lose contact with his natural mother. If he does he must quickly find another female willing to nurse him.

About three weeks after birth cows lose interest in their babies and focus their attention on the bulls. Now comes the mating frenzy. On Guadalupe we observed the old harem patterns completely broken down: all mating relationships became strictly for-the-moment; females mated with several bulls in the same day; bulls, winners and losers both, simply moved in whenever they spotted an opportunity. During quieter moments the huge bulls might embrace the females with their own kind of tenderness. But such peaceful interludes are brief. The females are fertile for only a few weeks—and the drive to mate creates a nightmare world, especially for the young. Approximately one-third of newborn pups die.

For the sea elephant every moment on land means a tremendous physical effort. Because of his size he is at the outer limits of his ability to survive out of water. He tires quickly; his body overheats; except when impelled by instinct to fight or mate he spends most of his land time in a stupor. In the nineteenth century this limitation led to the near extinction of the species, for against man on land the sea elephant stands no chance at all. Fifty years of slaughter by sealers and whalers reduced the vast herds of northern sea elephants to a remnant colony of less than 100 animals isolated on Guadalupe. But the sea—and the Mexican government—rescued them. Capable of diving 1000 feet deep and surfacing more than a mile away, in the sea the animal is practically invulnerable to man's attack. Sufficient animals were at sea to bolster the dwindling stock. Today what the sea elephant needs more than anything else is another isolated island—or *twenty!*

Quieter moments. *Because females are fertile for only a brief period each year, the drive for procreation creates an hysterical atmosphere. A male sea elephant and his harem are seen here in a rare, quiet moment on land.*

Man and grouper. *Diver Raymond Coll locating the grouper's house in order to place the mirror at the entrance and test the grouper's aggression.*

Protective grouper. *Encountering his own reflection, the grouper tried to scare away the "intruder." When all else failed, he was forced to battle.*

Territorial Behavior

In mating season the profligate codfish becomes territorial, with dominant males creating zones around themselves that cannot be entered by rival males or unripe females. The victor in battle is the animal who also passes along his strength to the new generation. How important his own particular kingdom is to the gentle, lethargic grouper is shown here. Within his territory the grouper had a house, a deep crevice in the coral. To this, *Calypso* diver Raymond Coll took an object a grouper in an Indian Ocean reef had never before seen, a mirror. The grouper reacted as we expected him to, by attempting to frighten the "intruder" away. Baffled by the response of his mirror image, the grouper tried a new tactic, to outflank the challenger. Behind the mirror there was no challenger, and the grouper returned to his house only to find his rival returned as well. To further observe his reactions Coll added three more mirrors, surrounding the house. As the grouper normally patrols all his borders, he found this situation too much to bear and proceeded to attack the fish in the mirror, biting it and challenging Coll as well. Finally, his head blows broke one of the mirrors into pieces. We had seen that in the last resort the grouper will fight.

The Defensive Tuskfish

Many territorial reef fish are conspicuously marked and display bright colors. Found on the Great Barrier Reef of Australia, the harlequin tuskfish seeks out and defends an area of the pockmarked reef. Its dramatic markings probably enable other fish of the same species to recognize it. Intruders usually avoid passing over occupied territory, but apparently one of the fish pictured chose not to.

Chapter IX. Chancelings

Many sea animals reproduce by laying eggs. Some species conceal or expose them, deposit them in nests or in the open sea, guard them or leave them on their own to hatch. To the latter I give the name *chancelings*. Their survival depends entirely on chance. For once spawned and hatched, they have mainly luck and instinct to assist them.

Adults of chanceling species are provided with the ability to lay large numbers of eggs, by sheer number giving their species the best survival chance. The oyster, depending upon its surroundings, is able to produce from 500,000 to 500 million eggs.

Of course, we must realize that with these fantastic numbers of eggs deposited there is an equally large number of eggs which do not hatch, or which do not survive to reach adolescence and maturity. The reasons for eggs not hatching are many. They are eaten by other types of life, marine and nonmarine, as they help to form the base of the food chain. In a single gulp a large creature can consume hundreds of thousands of eggs. The eggs may also be carried to inhospitable waters too cold for hatching, or tossed ashore where they will die in the hot sun. Other eggs may be covered by algae.

The offspring of most brittle starfish are chancelings, although some of the species protect their eggs until they are hatched. The various species of marine animals which do not care for their offspring, either by guarding the eggs or caring for the young after hatching, are protected from extinction by the ability to lay eggs in incredible numbers—in some cases billions per female. Predators find the undefended eggs and vulnerable hatchlings an easy and tasty meal: So, because only the smallest percentage of eggs reaches maturity, trillions of eggs must be deposited throughout the world's oceans, an abundance permitting a sufficient "hatch ratio" to ensure species survival. Those few hatchlings which reach maturity lay comparable numbers of eggs.

Despite the fact that no parental protection can be provided to chancelings during incubation and after hatching, some parents do make attempts to ensure a somewhat increased rate of survival among the eggs and hatchlings by depositing them in relatively protected places. These practices, evolved over millions of years, are often ingenious. The grunion and the green sea turtle deposit their eggs in the sand, above the high-water mark, affording protection from nest robbers of the deep, but, on the other hand, exposing eggs and hatchlings to land predators like coconut crabs or frigate birds. The Pacific salmon swims far up to the headwaters of freshwater streams to lay its eggs. In an area so far removed from the sea there are rela-

> **"The incredibly prolific oyster, depending upon its surroundings, is able to produce from 500,000 to 500 million eggs."**

tively few predators; the eggs have greatly increased chances for hatching. But by doing so the parents take new risks for themselves. Since a larger percentage of these carefully laid eggs hatch and survive the early phases of life, these species need deposit smaller numbers of eggs.

Brittle starfish surrounding a sponge. Both animals shed their eggs and sperm to the currents, ensured of reproductive success primarily by the sheer number of gametes released. The brittle stars wait with elevated arms to obtain planktonic food.

Bubble shell. *Above, this tentacled mollusc, rarely over one-half inch in length, is here seen laying eggs. Each bubble shell lays millions of eggs each year.*

Sea hare's egg masses. *This mollusc weighs an average of eight pounds. A sea hare has been known to lay 500 million eggs in less than five months!*

Prolific Producers

Each of the many thousands of species of plants and animals in the ocean differs in some way from all the others—in size and appearance, in temperament and behavior, in strength and longevity, in the way they court, mate, reproduce. The enormous diversity of species is nowhere more noticeable than it is in comparing the fecundity of marine animals. We have seen many instances of the incredible prolificacy of most marine life. On the other hand, human females and female ocean-dwelling mammals produce only a few eggs each year. Of these, rarely more than a single egg is fertilized.

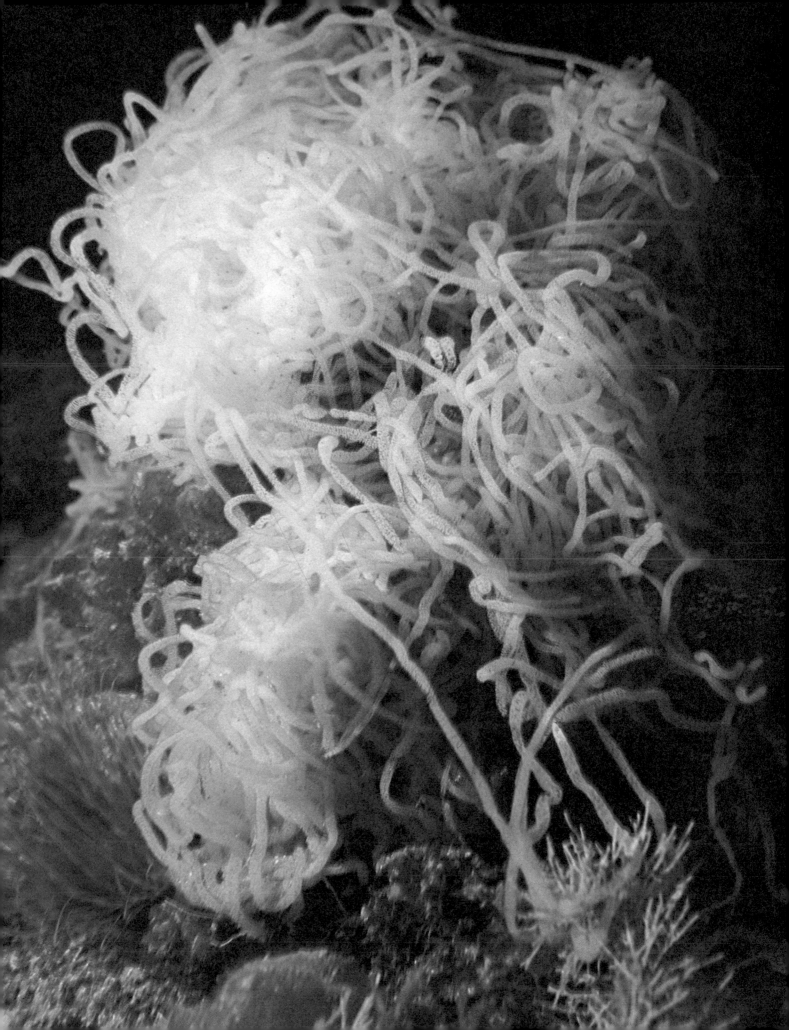

What If All Survived?

In spite of the apparent hit-or-miss methods of reproduction, and the odds against a fish's reaching maturity, populations in the oceans are relatively stable. What, we wonder, would happen if, by some quirk of fate, *all* the eggs of one of the prolific spawners were to survive and reach maturity? Let us start with the annual production of a single pair of codfish—9 million! The food provided in its yolk sac exhausted, the little fish begin eating plankton. In two years they will have grown to 15 inches, and in five years will be mature and ready to spawn. This generation could produce over 40 trillion offspring in the fifth year. One scientist has speculated that without "balances" it would take only a few years to pack the Atlantic from shore to shore, from surface to abyss, with codfish.

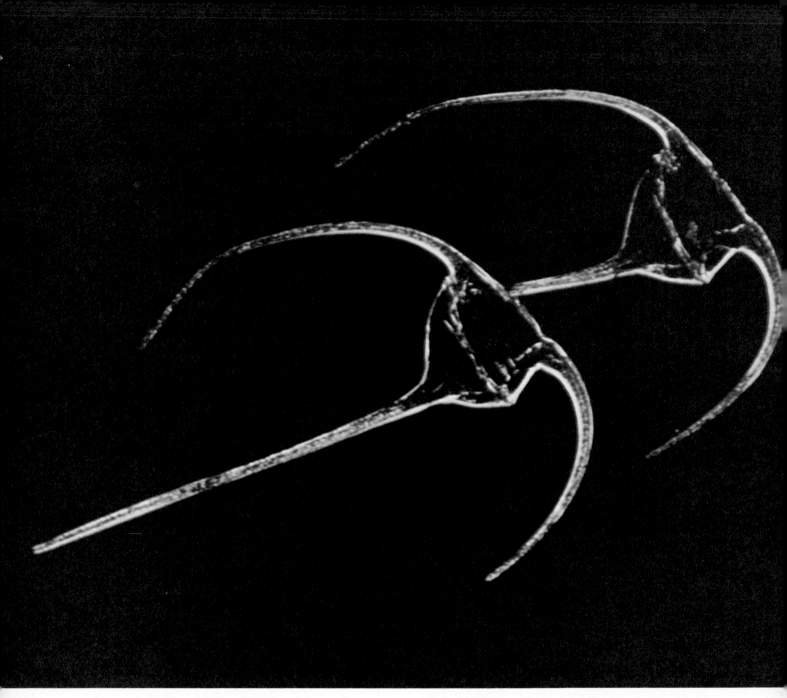

Dinoflagellates. *Dinoflagellates are one form of the drifting, microscopic plants of the sea. But a population explosion causes the death of huge numbers of marine species.*

The result. *The several hundred fish pictured here are but a portion of the animals wiped out by the "red tide." The dinoflagellates, too, end up decimated by their own overpopulation.*

Red Tide

Frequently our newspapers report alarming fishkills in the water off the coast of Florida. These disasters are thought to be the result of a population explosion of tiny organisms, dinoflagellates. When there is an increase in the nutrients near the surface, plus a stable layer of warm surface water, dinoflagellates multiply at a phenomenal rate. The name "red tide" derives from their color and a toxic pigment they secrete. It may be this poison that kills the fish, or it may be that the burgeoning dinoflagellate population robs the fish of oxygen. The dinoflagellate superpopulations soon collapse.

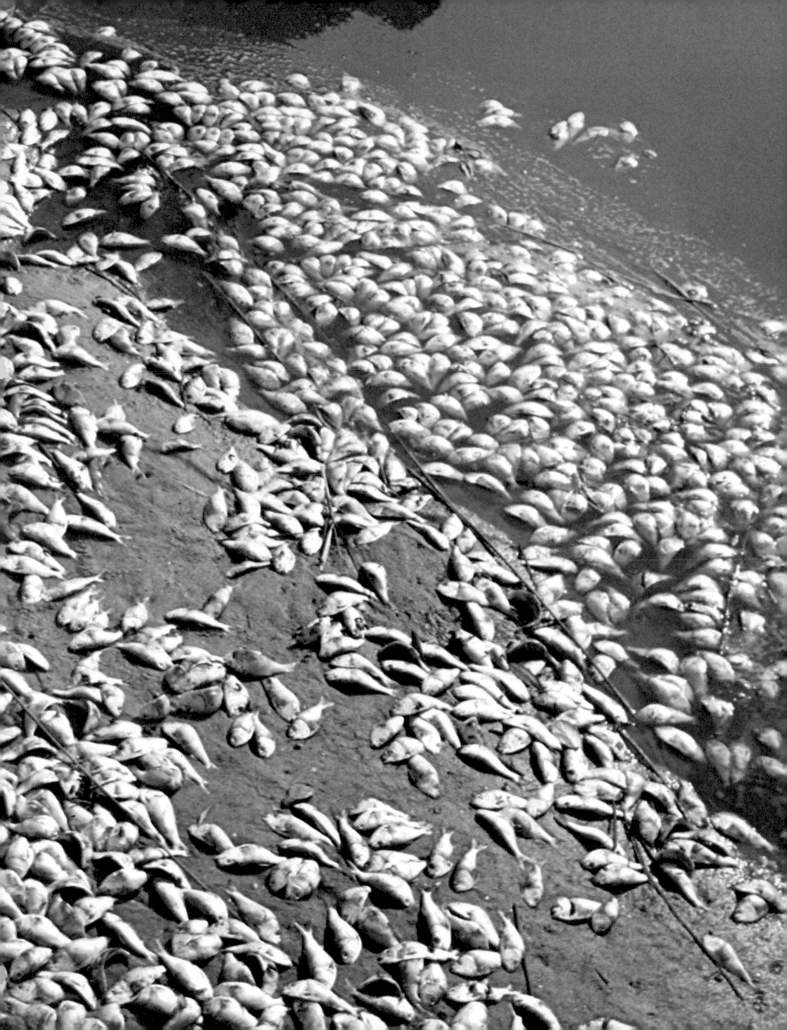

▲ A

Sheltering the Eggs

There are various manners in which marine species, from highest to lowest, shelter their eggs. Some enclose them in a gelatinous mass, others cement them to rocks, still others bury them at the edge of the sea. All of them, however, over the eons, have developed suitable means of making certain their species survive.

A / Green sea turtle. The female green sea turtle timidly pulls herself from her ocean home on dark summer nights and laboriously plods up sandy tropical beaches. Once on the beach the turtle, which has probably just mated offshore, carefully scoops a hole in the warm moist sand and, with tears streaming from her eyes (washing sand from them), deposits her leathery eggs. The excavated sand is replaced, covering the precious nest, before she makes her way back to the sea.

B / Nudibranch eggs. The nudibranch, a shell-less mollusc, covers its eggs only with water and mucus. For protection, her eggs contain a substance which makes them distasteful to predators, who carefully avoid an otherwise easy and rich meal.

C / Cone snail. The cone snail, another mollusc, like the nudibranch, does not need to dig nests or seek out secluded places, but is able to lay eggs in open, convenient places. Its eggs, too, have a distasteful outer covering which serves as protection.

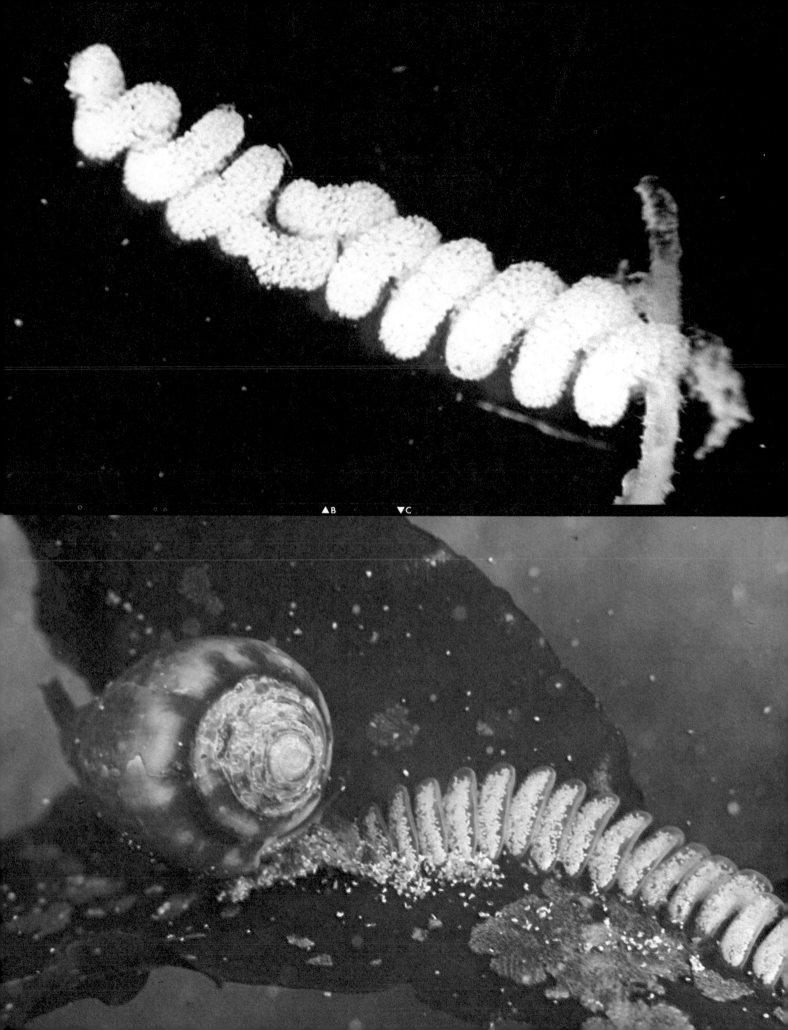

▲B ▼C

Protective Casings

Some marine animals create structures around their eggs to protect them during this vulnerable phase of life. On the previous page we saw methods of shelter and natural protection. Here we will see some of the methods used by sea creatures to make certain their eggs are well-protected.

A | Ratfish. Like the ratfish above, skates have egg cases that are flattened rectangular sheaths made of keratin, the horny material of our fingernails. These cases have tendrils on their four corners, two of which are coated with a gummy material and serve to anchor the case to some object on the sea bottom. Depending on the temperature of the surrounding water, the skate embryo remains in the case for a period of four and a half to 14 months. When the egg is sufficiently mature, the case opens.

B | Paper nautilus. This mollusc forms a shell from secretions of two specialized arms and in this structure deposits and hatches her eggs. The single-chambered shell also acts as a cover for the animal and as a hydrostatic organ, making her buoyant. She is not attached to the shell and can leave temporarily if she likes, but she will die if she permanently loses it.

C | Sea collar. The moon snail exudes a slimy mucus cementing her eggs together with grains of sand. When the mucus hardens to a leathery consistency, she slips out of the cape-shaped formation, which adheres to the bottom where the eggs will hatch. If the container is washed ashore it dries out, becoming brittle and fragile. Held up to bright light, the eggs appear as tiny, translucent dots in swirls of sand, often referred to as a "sea collar."

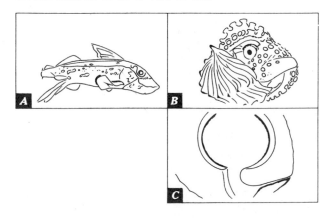

Nest Building

The stickleback presents us with one of the most dramatic examples of male aggression and domination of the mating ritual.

There are several stickleback species, distinguished by the number of spines along the back. And we need look only to familiar ponds and streams to witness the stickleback performing one of the most formal and grandiose nesting routines among fishes, signifying the onset of their breeding season. The male builds his roofed-over nest in serene waters among weeds. It's a very fine and carefully constructed nest, tubular in shape, made from small pieces of stems, roots, and fragments of sea plants. The stickleback's body is endowed with a sticky, gluelike secretion, emanating from his kidneys, which he uses to bind the structure together, shaping the nest to his liking by rubbing his body against the fragments much as a sculptor molds clay with his hands. Shortly the secretion hardens into a cementlike texture, resulting in a beautiful, durable nest.

> "He first tries
> gentle persuasion, including
> a courtship dance.
> But if that doesn't work
> he resorts to force.
> He will try to chase the female
> into his nest and if
> she is reluctant he hastens her
> along by nipping her tail."

His nest complete, the male now seeks a mating partner. Actually, he took the first step in this direction when he began to build

his nest. It was then that he dresses himself as handsomely as possible to attract the eyes of prospective females. The three-spined stickleback, normally blue or green with silver belly, bursts into a flaming red below. The ten-spined stickleback turns brown; the 15-spined, blue. He now begins to approach potential brides. If when doing so other males contest him, a ferocious fight breaks out. The goal is simple. He tries to coax a female to enter his nest, deposit two or three eggs, and promptly leave so he can get about his business. He first tries gentle persuasion, including a courtship dance, but if that doesn't work he resorts to force. He will try to chase the female into his nest and if she is reluctant he hastens her along by nipping her tail.

After the female deposits her eggs the male spews milt over them. By now the female

Stickleback. Ranging from one and one-half to four inches in length, sticklebacks feed on fish eggs, larvae, and small crustaceans. Fights between jealous males are frequent during spring and summer spawning.

should have departed. If not, he chases or pushes her out. Then he leaves the nest seeking another mate, repeating the process several times until his nest is full of eggs. When this is done he becomes all fatherly protector, guarding the eggs for a month or so against intruders. Should the nest become damaged, he promptly repairs it. He aerates the eggs by swimming around the nest, fanning with his pectoral fins. When they hatch he destroys all but the foundation of the nest. This he leaves as a cradle for the fry, which he continues to guard until they are ready to care for themselves. When that time comes the male simply swims away.

Mouthbreeders

The jawfish, which digs a 12- to 14-inch burrow with an enlarged internal chamber, gives birth in a remarkable way. At least two

> **"These fish have large mouths, useful at breeding time since they scoop their fertilized eggs into them to hold carefully until hatching."**

species are known to be mouthbreeders. These fish have large mouths, extremely useful at breeding time since these fish scoop

their fertilized eggs into them and carefully hold them there until hatching. Their mouths are not big enough to contain the egg masses when closed. So the jawfish keeps its mouth open for the incubation.

Other fish have also evolved this method of protecting their eggs. The mouthbreeding catfish and tilapia are among them. The male pipefish, cousin to the seahorse, protects his eggs by carrying them on his abdomen or under his tail, sometimes covered by a flap of skin. The obstetrical catfish of South America lays eggs which have special stalks by which the eggs attach themselves to the underside of the mother.

The Great Liberator

The garibaldi is a largish member of the damselfish family—coming in at about 11 inches in length, commonly found in the temperate waters of the kelp-bed ecosystem off southern California. His brilliant orange color, plus his combativeness, explains his name. The garibaldi is the embodiment of the territoriality principle. He will not hesitate to remove objects twice his size from his nest. In this picture we see one of these belligerent little animals reacting to the practical joke of one of our divers

> "The garibaldi is the embodiment of the territoriality principle. He will not hesitate to remove objects twice his size from his nest."

who placed a starfish close to his nest. The garibaldi promptly moved in, picked up the intruder, and showed him the gate.

The Great Protector

The female garibaldi lays her eggs on algae, the male cleans an area around them. And he guards this area fiercely. In the grip of the territorial urge this animal seems to turn indifferent to the size of the intruder. This behavior almost doomed them as a common inshore species because, as they stood their ground uttering thumping sounds as a threat, even the worst spearfishermen could hardly miss the bright orange target.

"In the grip of
the territorial urge
this animal seems
to turn indifferent to
the size of the intruder."

The picture above shows the nest closeup. The eggs are attached to red algae, which is hardly visible. Within the ready-to-hatch eggs, young can be identified by the silvery reflection from their tiny paired eyes.

Upwelling

FISHING GROUNDS

DENSE POPULATION

The sea's plant and animal populations are unevenly distributed. Vast expanses of open ocean are deserted. The density of sea life depends upon two main ingredients required for photosynthesis: sunlight and nutrients. Sunlight penetrates only a few hundred feet. Thus, organic material that settles to the

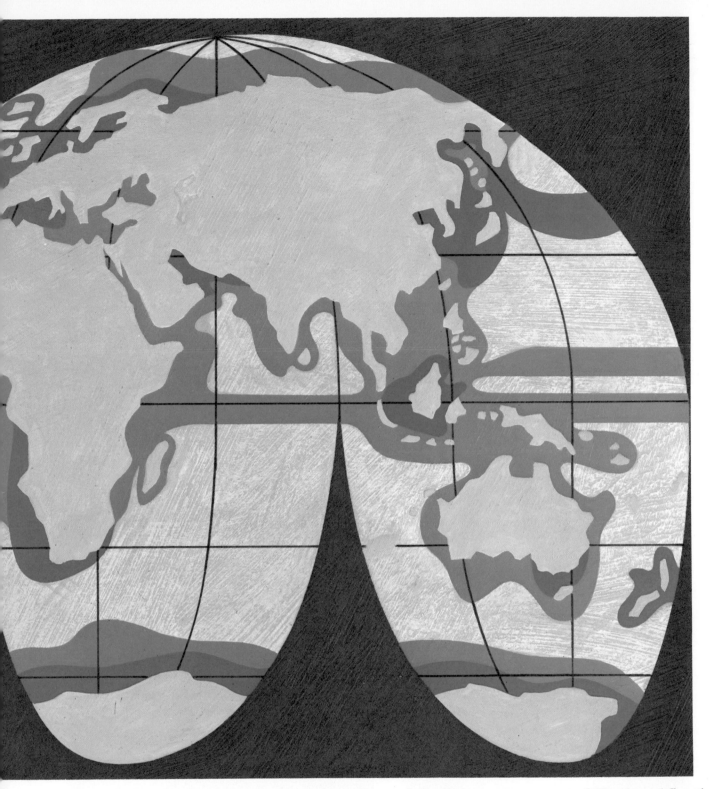

floor of the ocean must be returned to the sunlit layer or it is lost to the life cycle. This is accomplished by "upwelling," vertical migration of water caused by winds, currents, or density differences of the water. Predictably, large populations are found in areas where upwelling is regular or continuous, as it is off the western coast of North and South America, southwest Africa, and western Australia. Most of the world's commercial fishing fleet operates in regions where upwelling furnishes regenerated nutrients to the basis of the food chain, the meadows of phytoplankton.

Specially Timed Spawning

A phenomenon occurs every spring and summer along the warm sandy beaches of southern California. During the months from March through September the beaches are invaded by seven-inch-long silvery fishes coming ashore from the sea. These smeltlike fish, called grunions, are a source of infinite delight to fish watchers.

The grunion spawns during these months only on the three or four nights following the new or full moon. Then the spring tides are receding from their highest, and the water tosses the little fish far up on the beaches, above the normal high-water mark, where the sand is warmed by the sun and

> "A female grunion makes her run generally between 9 and midnight under cover of darkness, fairly safe from hungry sea-birds. By laying eggs in this manner, she offers her offspring a form of parental care, as at least she has protected them from sea predators."

moistened, after having been touched by the sea. Finding herself in such a spot, the female grunion wriggles frantically on her tail, drilling a hole in the sand. When approximately two-thirds of her length is buried, with only head and pectoral fins exposed, from one to as many as ten males arch themselves around her body. The female then begins to deposit eggs in her sandy hole, the males to discharge milt that runs down her body reaching and fertilizing the eggs. When her eggs (she may lay from 1000 to 3000 every two weeks) have been safely deposited and fertilized, she leaves the hole and makes her way back to the sea. The outgoing tide deposits more sand on the beach, covering the eggs with eight to 16 inches of sand.

Approximately two weeks later, the sun, moon, and earth in line once more, high tides wash the sand from atop the nest and stimulate the tiny grunion larvae to pop from their eggs and make their way seaward. The embryos develop sufficiently to hatch in from one week to ten days, but they wait for the high tide which will wash them out to sea. The grunion eggs are unusual in another way. Should the unhatched, mature eggs fail to be uncovered and agitated by the surf, they can remain buried for an additional two weeks until the next high tide and hatch with no ill effects. The process repeats itself every two weeks or so for the several summer months, with the female making many trips out of the sea and up onto the beaches. She makes her run generally between nine and midnight under cover of darkness, fairly safe from hungry seabirds. By laying eggs in this manner, she offers her offspring a form of parental care.

How many of the litter survive is an unknown factor. Grunion mature sexually in a single year. Curiously, these rapidly maturing fish must decline rapidly sexually. Only 25 percent of grunion lay eggs at the end of their second year, only 7 percent lay eggs at the end of their third year. By the end of the fourth year of life, no grunion lay eggs. Grunions are easily picked up by observers at their egg-laying ritual, as they are quite awkward on land. The little fish, once on the brink of extinction, is now protected by law.

Grunions. These fish have a limited period of sexual activity. Only 25 percent of grunions lay eggs at the end of the second year, only 7 percent at the end of their third year. By the end of the fourth year of life no grunions lay eggs.

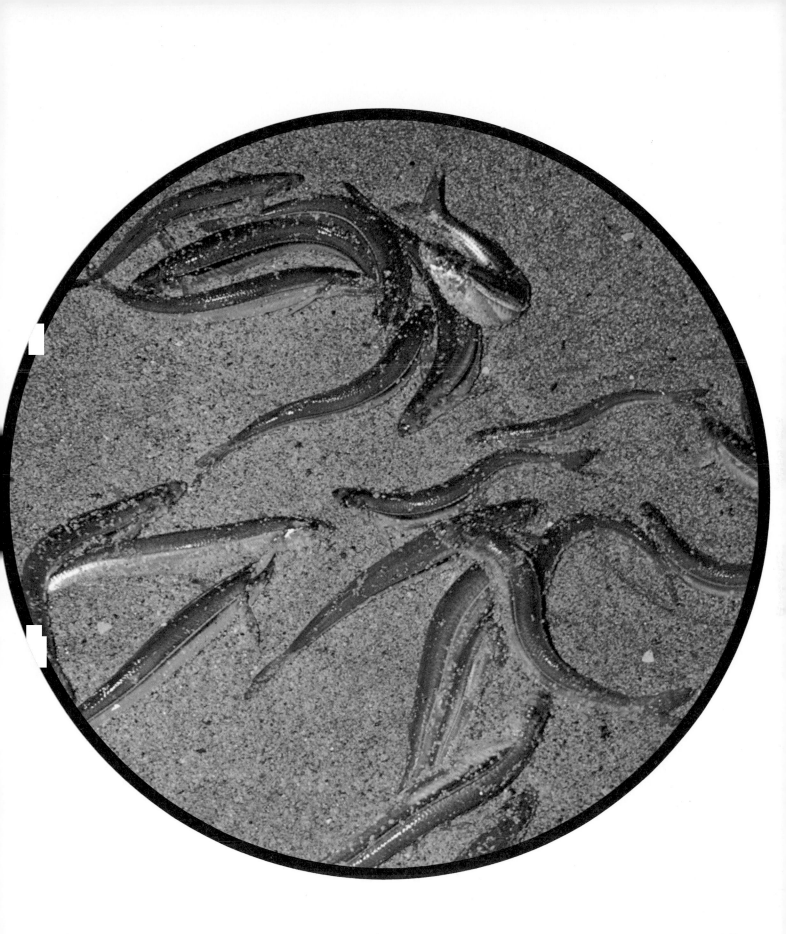

Icicles melting. As the sun gets higher in the sky with the changing seasons, even polar ice begins to melt—and a small summer begins.

The Seasons in the Ocean

In all but equatorial seas the seasons in the oceans are characterized by population fluctuations. Sea plants are quiescent in winter. They require warm sunlight for photosynthesis. But the storms of winter supply another of the vegetation's requirements. They stir the water with sediments, making it a nutrient-rich broth. When the sun moves into higher latitudes, the algae explode into bloom. Larval forms join in the planktonic layers, grazing on the plants and depleting them as the plants depleted the dissolved phosphates and nitrates. Larger animals feed on smaller ones, decreasing their numbers. Loss of populations at the base of the food chain continues until early autumn when

108

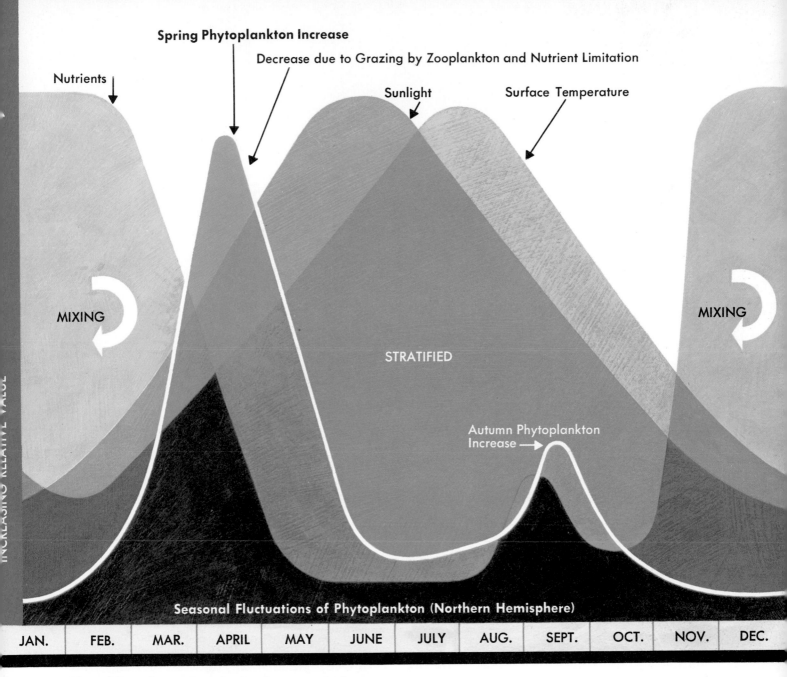

Seasonal Fluctuations of Phytoplankton (Northern Hemisphere)

| JAN. | FEB. | MAR. | APRIL | MAY | JUNE | JULY | AUG. | SEPT. | OCT. | NOV. | DEC. |

Variable values. Changes in the amount of sunlight and water temperatures throughout the year have varying effects on planktonic populations.

the surface cools and storm waves return some nutrients to the surface. The renaissance is short. The sunlight soon weakens.

In tropic waters seasons have little effect on productivity and reproduction. The fluctuations of populations are minor. The most spectacular springtime is in the polar seas, where melting ice sets up a vertical migration of water and the upwelling stimulates a dramatic multiplication of diatoms. This in turn invites a similar explosion of krill, small shrimplike animals that can carpet more than 3500 square miles of sea surface. The multitudinous krill comprise the main food supply for the great baleen whales.

Puget Sound, Washington. *Above, man has fenced in open water in order to farm salmon, much as land farmers till their soil.*

Man-made pens. *Man can also construct closed-in pens to protect the vulnerable hatchlings until they are ready to go to sea.*

The Need for Mariculture

In his exploitation of the sea man is still a barbarian, a ruthless hunter slaughtering whole species of animals without heeding the consequences. With earth's burgeoning human populations to feed we must turn to the sea with new understanding and new technology. We need to farm it as we farm the land. This is called mariculture. It has just begun.

We know that life in the sea depends upon the sunlight and the nutrients necessary for phytoplankton. Vast stretches of the sea are desert because these two elements do not coincide. There are three ways in which we can improve the situation. Nutrient-rich deep water could be pumped to the surface in artificial upwellings; light energy could be introduced into the depths; and fertilizers could be dispersed over the surface.

But mariculture can be less ambitious and start with properly managing limited bodies of water. In such controlled volumes the ideal conditions can be maintained all year, and by ensuring fertilization and protecting the larvae from predators, incredibly high yields can be obtained from a number of protein-rich populations. High-efficiency sea farms totaling the size of Switzerland would produce more food than all fisheries combined.

Adult wolf eels with eggs. *Above, adult wolf eels are seen with their eggs. They wrap their bodies around the eggs to protect them.*

Larval wolf eels. *Parental care goes just so far, however. Once the eggs are hatched the larval wolf eels are on their own and must protect themselves.*

Standing Guard Over Eggs

After hatching from their buoyant eggs the young of most marine animals, whether pelagic or bottom-dwelling, live and feed for a period of time near the sea's surface—as part of the plankton. In many cases trans- formation from larvae into adults occurs rapidly. In other fairly rare cases the process may take up to several years. Life in the plankton has its advantages—and its drawbacks. It provides larvae with a rich, readily available food source, but it also exposes them as a food to larger animals.

The Role of the Predator

In the sea, as we have seen, there is a limited amount of room and stock of materials necessary to life. Flesh and bones are built by organic materials circulating constantly through the food chain. At one time most of them will be in vegetation, then they will be absorbed by animals, then by other animals, and they will finally settle as detritus and decaying flesh to the floor of the sea where they will be converted by bacteria into useful nutrients to begin again. Some of them are lost, absorbed by the thick bottom mud. The land itself contributes through rivers that drain the continents. The quantity of nutrients available varies only slightly.

The drive to reproduce is universal in plants and animals. If we assume that the population in the world's oceans at this time is in balance, with the desirable numbers of each

mature plant and animal flourishing, it is obvious that each adult male and female needs only have two offspring that survive and reproduce in their turn. This is zero population growth.

The predator has been miscast in the role of a villain, for predation is needed to ensure stable, healthy communities. Diving scientists recently discovered a reef that had lost its predators to overfishing. The remaining animals were sickly and some were unable

Angelfish. *These beautifully colored fish move gracefully in and around tropical waters across the globe. With diets consisting of crabs, barnacles, and other invertebrates, these fish can grow to two feet.*

to swim upright. The usefulness of predators in maintaining healthy populations of prey has been demonstrated on land.

Man is the only animal with the ability to change the environment he lives in, defy natural selection, and exist outside the rules. For how long?

Chapter X. Internal Parental Protection

Dolphins, whales, and humans are all viviparous—they give birth to living young. Viviparous females produce a few eggs, one or more of which is fertilized by the male while still inside the female. As the animal develops through its embryonic stages in the uterus, it is like a parasite of the female—attached by an umbilical cord through which food and oxygen travel from female to offspring and waste materials from offspring to female, the female sustaining the offspring by means of her own bloodstream. Not all viviparous animals in the sea are mammals. Many species of sharks and some bony fish give birth to live offspring.

Another experiment undertaken by nature occurs in the ovoviviparous animals. These species, including some sharks and manta rays, use a method of reproduction which may be considered midway between the egg-laying oviparous animals (such as birds) and viviparous animals like dolphins and man. The female of an ovoviviparous animal also releases an egg that is internally fertilized by the male, but it is retained in the oviduct. Lacking an actual uterus, the oviduct becomes a sort of womb.

Another theory holds that projections developing on the "uterine" walls may grow into the digestive tract of the offspring, secreting nutritive fluids that provide nourishment for the embryo. By being allowed to mature and hatch internally, eggs are protected from the hazards faced by eggs which are simply dumped in the sea.

The seahorse female plays only the briefest role in the prenatal care and delivery of the next generation. It is a strange reproductive procedure. The female produces the eggs and deposits them in a minute opening in the male's brood pouch located under his tail.

As she swims away, never again to see her offspring, it becomes the male's responsibility to fertilize and incubate the eggs as well as to deliver the newborn. Gradually his pouch swells as the eggs grow bigger and capillary blood vessels multiply. A form of protective tissue develops around each egg, ensuring it enough room inside the pouch. The incubation period lasts from eight to

> "The female seahorse produces the eggs and immediately deposits them in a minute opening in the male's brood pouch located under its tail. As she swims away, it becomes the male's responsibility to fertilize and incubate the eggs as well as to deliver the newborn."

ten days. When the male is ready to give birth, his body begins to move back and forth, tensing at intervals, as he thrusts forward his bulging pouch. Slowly the pouch opens, and with continued rhythmic tensions and convulsive jerks he begins to eject his offspring, usually one at a time. Such "labor pains" continue for several hours after which several dozen to many hundreds of baby seahorses have been born, depending on the species. Having performed this birth function, the male seahorse leaves the babies to care for themselves, and even gulps a few. They at once begin swimming in search of suitable blades of grass around which they can twist their tails and begin growth.

Male seahorse and offspring. In the photograph at right, the male seahorse has just given birth to many tiny seahorses, seen floating in the water above him. After this task is completed, however, there is no parental care—the baby seahorses must fend for themselves.

Shark being born. Some sharks, like the one pictured here, give birth to live young. Others lay eggs that are later hatched.

Advanced Breeding

The mating practices of sharks and rays are among the most advanced of all the fishes. The female carries eggs inside her body. The male is equipped with a pair of erectible rods called claspers. Through these claspers he releases his sperm into the female's body opening. Sometime after fertilization certain sharks and skates release the eggs directly into the sea. However, most open-sea sharks and rays are live-bearers, producing fully formed offspring at birth—pups who have experienced complete embryonic development inside the female's oviduct prior to birth and who during that period were nourished by the female and from the yolk of the large yolk sac surrounding them. As for rays, it is believed that the manta gives birth to a single pup. During the birth process and often at other times, the manta flaps its fins much as a bird does its wings and leaps as high as five or six feet out of the sea.

The actual birth. *In this remarkable photograph we are watching a dolphin being born, tail first. The baby dolphin is usually about three and one-half feet in length, and reaches sexual maturity at around four and one-half years of age.*

Birth of a Dolphin

After about half of the dolphin's 11- to 12-month gestation period the pregnant female isolates herself slightly from the herd, concerned with the coming ordeal. She may choose a single female companion to keep her company, to stand by and assist her. As the weeks go by, the female begins what appear to be prenatal exercises, flexing her tail up and down, possibly to tone her muscles. Eventually the day arrives! The tail of the infant protrudes from the female's body. Gradually more and more of him emerges. The companion the female has chosen stands by ready to help, as do other females —observing the birth, keeping curious males away. When the infant is fully expelled from her body, the female whirls around, breaking the umbilical cord. She can now assist the young dolphin reach the surface for its first breath, and join the herd.

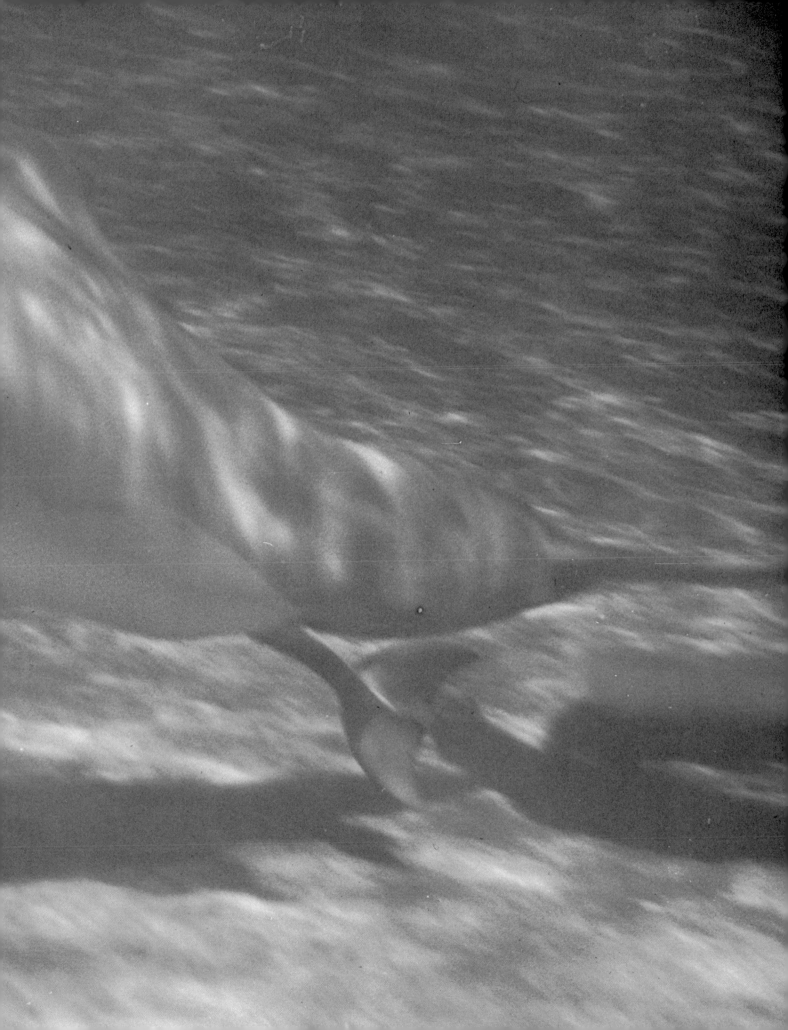

__Hours after birth.__ After the female dolphin has taken her baby to the surface for its first breath of air, they return and swim together. She stays with her offspring, caring and teaching, for over a year.

Similar Beginnings

Each a master of a world, man and dolphin have similar beginnings. During the early weeks of development the embryos are virtually indistinguishable. After conception the embryo of each remains within the mother, enclosed in the fluid-filled amniotic sac until, many months later, it has sufficiently matured to meet the challenges of life. While in the female's body each embryo is insulated against heat and cold. Each first receives nourishment from the attached yolk sac and later directly from the female. Growing steadily, its liquid environment cushioning it from shocks and bumps, a several-weeks-old embryo is large enough to be seen by the unaided eye but will not be fully formed for months to come. The fine vestigal hair which covers human fetuses and is found on some species of dolphins will nearly completely disappear before birth, when the bare-skinned infant makes his entry into the world he is destined to play in and rule.

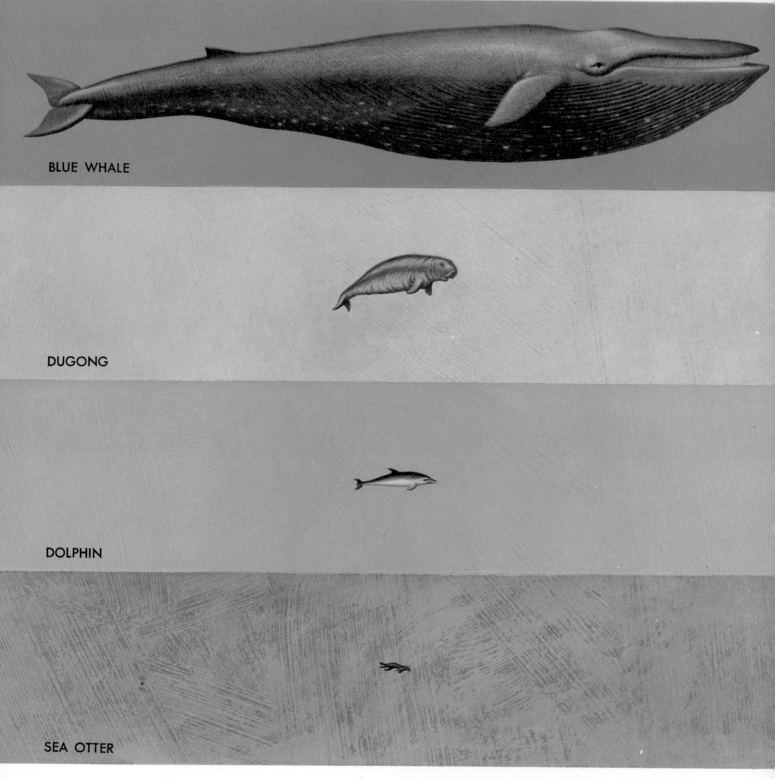

BLUE WHALE

DUGONG

DOLPHIN

SEA OTTER

Gestation and Weaning Periods

Despite enormous size differences, the gestation periods in mammals—the time it takes an offspring to mature from conception to birth—are remarkably similar. In the pro-

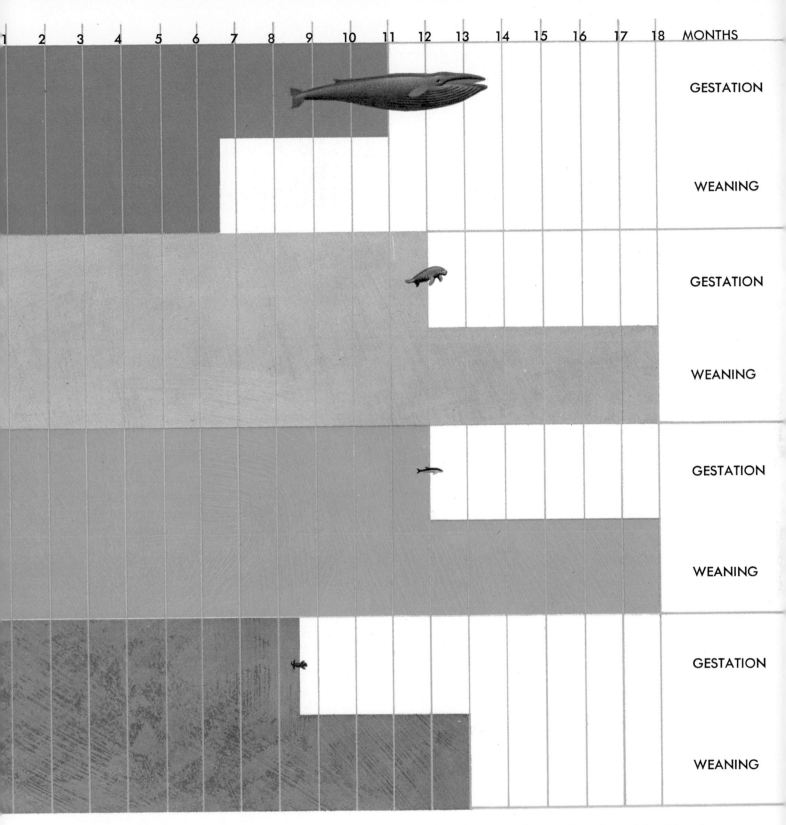

| 1 | 2 | 3 | 4 | 5 | 6 | 7 | 8 | 9 | 10 | 11 | 12 | 13 | 14 | 15 | 16 | 17 | 18 | MONTHS |

GESTATION

WEANING

GESTATION

WEANING

GESTATION

WEANING

GESTATION

WEANING

tective custody of the female's womb, the embryo is given full protection against the hazards that ravage the eggs spawned by lower forms. It takes all the nourishment it needs directly from the female's bloodstream. It need not grow at an accelerated rate, or undergo a sudden metamorphosis of form, in order to survive.

Chapter XI. Parental Care for the Young

We are often moved by the warmth, love, tenderness, care usually given the human child. This devotion is not unique.

As we probe into the concept of parental care, we begin to see patterns developing. The species which offer this care after birth do not produce offspring in great numbers. Correlated to a small fecundity is the relatively high survival rate of those few young. Parental care is extremely useful in improving the odds of the infant's survival during the early stages of life. During this period the offspring is susceptible to injury or death. The parent, or parents, serve as guards, protecting the young from danger, keeping them from blundering into harm. In any case, whether they produce millions of chancelings or a few carefully tended offspring, those species inadequate for actual living conditions are eliminated, and those successful species attempting to conquer the world are curtailed by the very destructions they produce. Man falls in this category.

In many cases the parent seeks out food for the young, sometimes partially digesting it. In other cases the offspring are led to the food, eliminating the perilous groping of the newly hatched young. Although the young of many fish can rely for a time on their yolk sac for sustenance, many die seeking that all-important first meal. In other situations the female herself is the source of food. In mammalian species the offspring live on milk produced by the female until their digestive systems or teeth have developed. Mammalian offspring, therefore, are given the opportunity to develop other skills before having to learn to forage for food.

Both female and male manatee take an interest in the upbringing of the young. Packing 400 pounds into her seven-foot, torpedo-like form, the female manatee may not look like one of the most affectionate animals in the sea, but she is. In the period after birth the 60-pound calf is held almost continuously. When the female is off chomping on her daily 100 pounds of vegetables, the male cradles the infant in his flippers. When the cow is finished, the bull manatee passes along the calf for *his* dinner, which the little one finds in nipples tucked in the folds of the female's skin at the base of each flipper. The females probably breed only every other year and keep their calves by them for near-

> "Those successful species attempting to conquer the world are curtailed by the very destructions they produce. Man falls in this category."

ly two years. Attempts are now being made to breed them and increase the dwindling population in Florida coastal waters.

Here is perhaps the most important advantage of parental care of offspring. The young of species who are cared for are given opportunities to observe the parent and to learn efficient methods of locating food and self-defense. This capacity for teaching and learning is what probably sets animals that care for their offspring apart from those that do not. The former are able to make use of what has gone before them and to make refinements in life practices.

Parent and offspring. Both manatee parents have equal interest in the upbringing of their offspring. Since an adult manatee spends about one-fourth of its day eating, the other parent must care for the offspring. They are found in tropical waters; the female gives birth in April or May. Nursing, which lasts as long as two years, can take place either submerged or on the surface.

Parade of the Whale

There is a marked parade order as the humpback whales progress on their yearly voyage from feeding grounds to breeding grounds and back. On the way to warm waters, females and their weaned yearlings are in the lead; the females will mate again this season. They are followed by adolescents and then by adult males and resting females. The pregnant females are last.

The humpback female gives her calf affection and attention, nursing the infant for more than a full year. On the long migration

> "At the approach of danger the female unhesitatingly positions herself between her offspring and the threat."

the female may take the infant under her fin and guide it along, or she may gently

push it with her rostrum. At the approach of danger, the female unhesitatingly positions herself between her offspring and the threat. Here she will be better able to fend off an attack and protect her baby from harm. Should the offspring be injured, the female supports it near the water's surface until it regains sufficient strength to breathe and swim again on its own. The humpback, like its dolphin relatives, has been known to continue supporting a dead infant for some days, even in the case of a stillborn which has never shown signs of life. The Greek philosopher and naturalist Aristotle concluded his study of whales by noting, "The creature is remarkable for the strength of its parental affection."

The affectionate whale. *A female humpback whale is here seen with her offspring. If the baby is hurt in any way, she stays with it near the surface until it is well enough to swim on alone.*

An Irresistible Mother

It would not be easy to find a more beautiful example in the animal world of "parental care for the young" than the sea otter. It spends all or nearly all its life at sea—indeed, giving birth in the sea. The social organization of the sea otter seems to be a

loose one, but there is an extraordinarily close relationship between female and offspring. After the birth the female carries her pup on her back as she paddles along, or hugs him to her breast. The pup is precocious and has a woolly coat at birth. When its mother is diving for food, the pup can float on its back even before it can swim on its belly. The pup, which has clawed toes on each foot helping him hold his prey, begins eating after it is only a few weeks old.

During the last century a single sea otter pelt could bring over $1000 on the European fur market, accounting for the supposed extinction of the animal by the year 1900. But a few individuals somehow survived, and now, under strict protection, the species is making a comeback. To have lost it would have been a grievous loss. The sea otter is one of the very, very few animals that can use tools: he has learned how to place a small stone on his chest as an anvil against which he breaks open abalone or other shellfish. He has also devised a way to anchor for the night. When he wants to sleep along a coastline somewhere, without being swept in or out by the tides, he gathers strands of kelp around his waist and nods off—securely moored.

Chapter XII. Perversions

Recent experiments performed by the pioneering American scientist John Calhoun have shown that overcrowding in rat communities causes breakdowns in normal patterns of behavior, resulting in what he calls "behavioral sinks." The sexual interests of some males change; they become hyperactive or homosexual or both. Other males become extremely phlegmatic. Females lose interest in their nests and offspring. Neglected offspring fail to survive.

In the artificial environment of an aquarium tank, and perhaps even in the open sea, males separated from females and unable to perform the sex act for long periods can react abnormally to stimuli, make advances toward substitute objects—a male of the same or other species, a female of a totally unrelated species.

The periodic mass migration of Norwegian lemmings—once thought to be searching

> "Males separated from females
> and unable to perform
> the sex act for long periods
> can react abnormally to stimuli,
> make advances toward
> substitute objects—a male
> of the same or other species,
> a female of a
> totally unrelated species."

for the lost continent of Atlantis—is a more final way of solving overpopulation problems. Those lemmings which don't commit suicide by jumping into the sea remain to reproduce and in several years are able to bring the population to numbers great enough to trigger another migration.

Wolf eels, like the rats in the Calhoun experiment, have been known to indulge in canni-

balism. Sometimes the female wolf eel, after carefully guarding her precious six-inch ball of eggs, calmly snaps up the hatchlings as they wriggle past her after hatching.

Examples of "perversions" in nature must be considered independently from human moral standards, but not from human social behavior. Almost extinct in 1911, when the Mexican government extended its protection, there has been a spectacular recovery in the sea elephant population of Guadalupe.

When I visited Guadalupe with my group we were immediately struck by the overwhelming sense of crowding in the rookeries—thousands of these giant creatures packed into a narrow area only a few hundred yards in length. The highly ordered harem society characteristic of this animal, which has been often observed with smaller groups, seemed to have been blown apart by the population explosion. No "territories" could be maintained. The great mammals seemed to fight whatever was next to them—or to make love to it.

So long as we are careful not to strike a moral note, behavior of this kind can be termed a "perversion"—or in the case of animals a deviation of instinct—because it is a behavior which seems to be biologically useless or even self-destructive. What seems a "right" or "wrong" activity frequently turns out to be either a meaningless activity or one which man simply does not sufficiently understand. We still are reluctant to admit that order is occasionally, but not inevitably, the product of chaos.

Overcrowding. Fur seals gather in huge numbers—sometimes hundreds of thousands. They have long been the victims of seamen, who hunt them for their oil, said to be better for lubricating than whale oil.

132

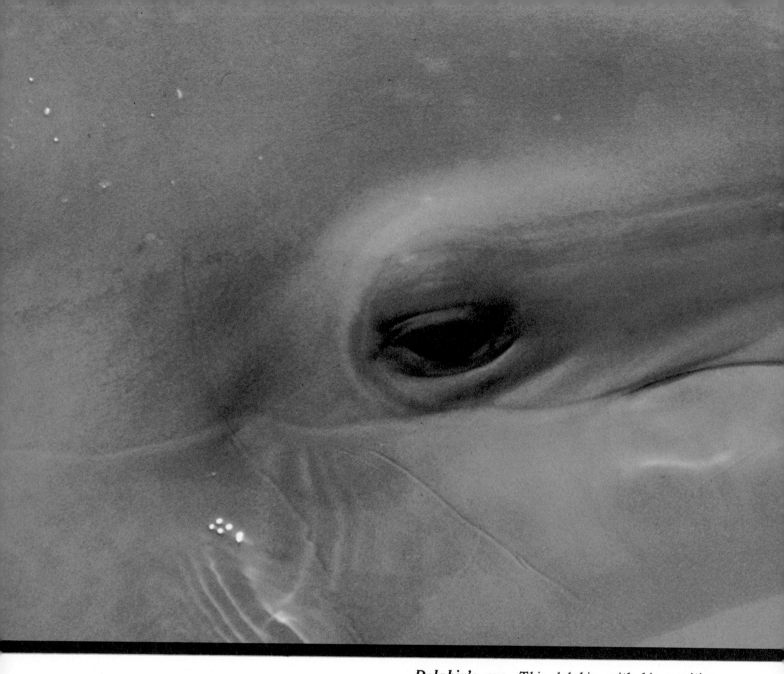

The Stress of Captivity

A popular attraction at the California Academy of Sciences' aquarium is the large dolphin tank. Along one side of the tank a stair leads down to a glass wall, enabling the spectator to see into the dolphin's underwater cage. A few years ago it was noticed that one of the bigger dolphins occasionally bled slightly from his intestine.

Now what? A dolphin can't be psychoanalyzed. Nor would it probably be easy to prescribe an ulcer diet for him, or to induce him to eat it were it prescribed. Working from the human premise that most ulcers are the results of "anxiety"—probably the only premise they had to go on—the worried doctors tried to find out what if anything was causing anxiety in their big favorite and how they might relieve it. At last they observed

Testing captive dolphin. Many tests have been made on dolphins in captivity. Here a dolphin's eyes were covered to try to discover how accurate his sonar really was. It proved surprisingly precise.

that the condition cleared up as soon as a shade was drawn across the glass wall separating visitors from the animals. All those human eyes staring into his privacy, admiring as the onlookers were, had apparently upset this one dolphin, making him nervous and perhaps resentful.

If this sort of reaction can take place against the accommodating backdrop of the California Academy installation, imagine the mental and physical distress of animals in less scrupulously maintained captivities. Majestic lions stalking back and forth in 20-foot concrete yards; an ape family given a couple of artificial rocks to sit on; a polar bear enduring a pitiless New York summer; a gigantic white rhinoceros with no mud wallow to roll in: the stresses of captivity often occasion breakdowns in the animal on display.

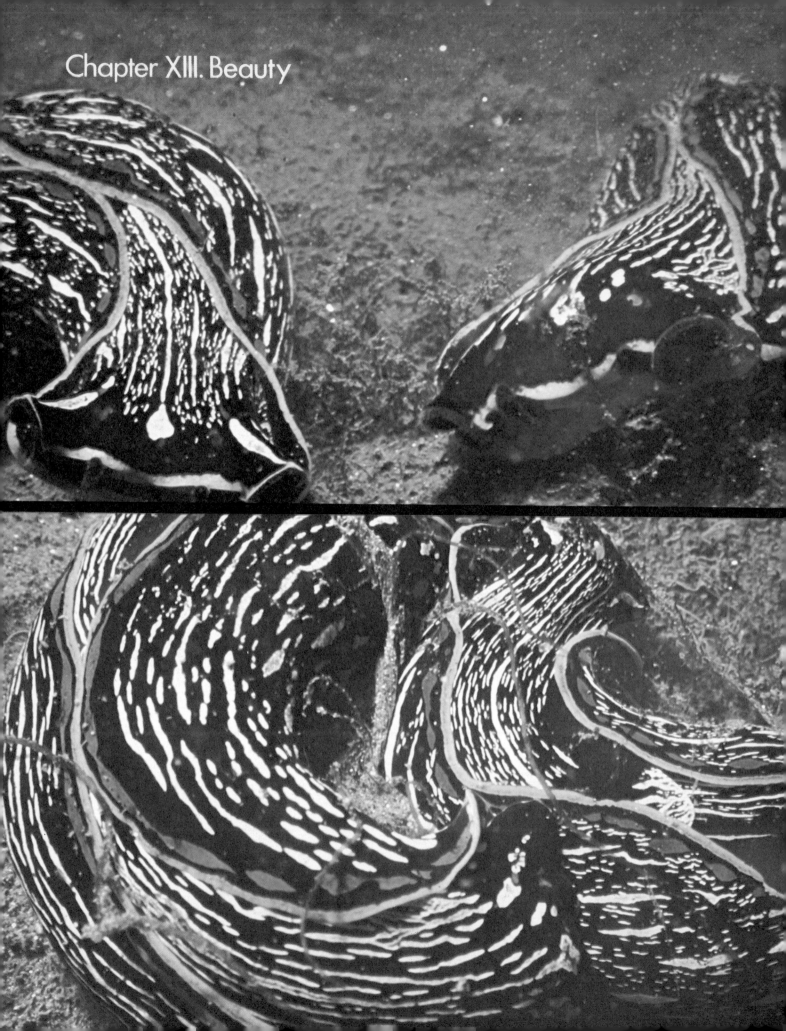

After chemically signaling their desire to mate, two young beautiful sea slugs, *Navanax inermis,* have come together out of their usual solitude in the shallow waters off the coast of southern California. They spend about an hour together with their supple bodies wrapped around each other, exchanging sperm. Then they separate, each to lay its eggs alone. The love scene occurs often in the life of the *Navanax,* continuing to the time of its death at the age of about eight months.

Jewels of the Sea

Among the most beautiful creatures in the world are those that dwell in the sea. A myriad of remarkable forms and colors take root, grow, or swim free. Even within a single species the different stages through which it passes on its way to sexual maturity can bring forth a spectacular array of shapes and hues. Pictured here are only a few of the treasures which may be found from polar to tropical waters. Spiraling out of the mother's body, ribbons of little eggs form coils and flowers. These grainy rosettes will not wither; in about two weeks new life will explode from them. Little anemones seem to bud from the parent, but they have been born inside the mother. They stop for a while to complete their development on her skin, then move off to begin an independent life.

A / Cone snail laying eggs. *These largely tropical poisonous molluscs lay their eggs in a gelatinous sheet that protects them.*

B / Budding anemone with turban snail. *Budding is one method by which anemones reproduce. The buds will break off after a while and be capable of reproducing by budding or egg laying or both.*

C / Nudibranch with ribbon of eggs. *Nudibranchs, a type of shell-less molluscs, are among the sea's most beautiful creatures. Even the manner in which the eggs are laid seems design-minded.*

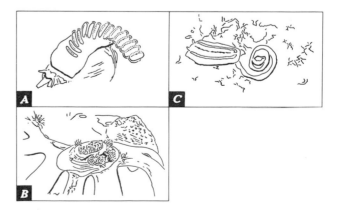

139

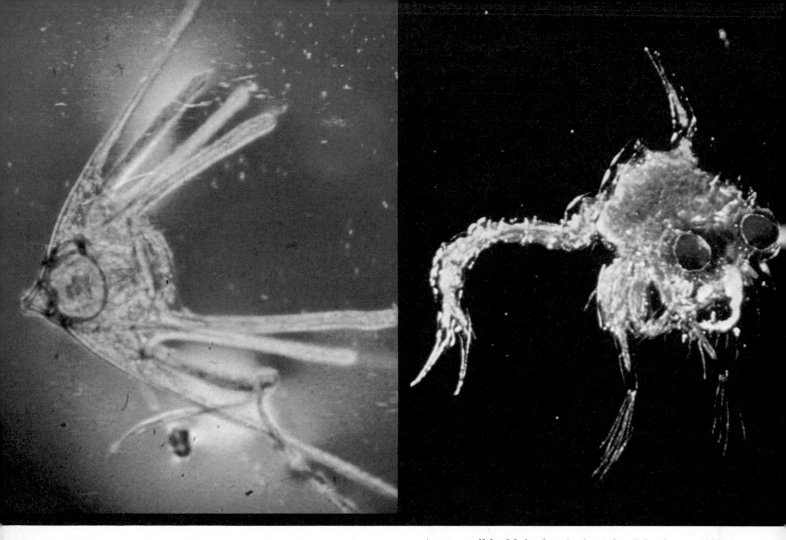

Hatchlings of the Sea

Seen through a microscope, the sea's tiny hatchlings gleam like jewels in a showcase. Prominent eyes glow from symmetrical transparent bodies. In larval form most of the little animals bear small resemblance to the adults that spawned them. Some will pass through several metamorphoses before leaving the planktonic stage. Each day of their young lives is filled with growing and learning—to move, to eat, to escape capture, to grow to maturity, somehow to survive the army of predators in order to give birth in their turn.

A / Brittle star larva. This beautiful creature will grow up to be from two to six inches in length with thin, flexible arms.

B / Crab larva. Seen here in one of its larval stages, the zoea stage, this crab will eventually grow to a possible 20 inches in length with pincers. The pincers are already seen forming here.

C / Sea urchin larva. This pluteus larve will develop into the purple-tipped sea urchin.

D / Crab larva. When it is full grown this will be an anomura—a variety of walking crab which may or may not be able to swim.

E / Starfish larva. A sinuous larval body and the larval tentacles give rise to the developing starfish, seen here as a partial orange mass.

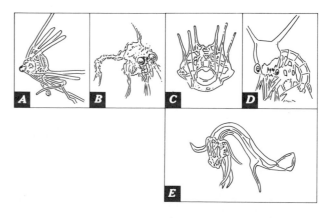

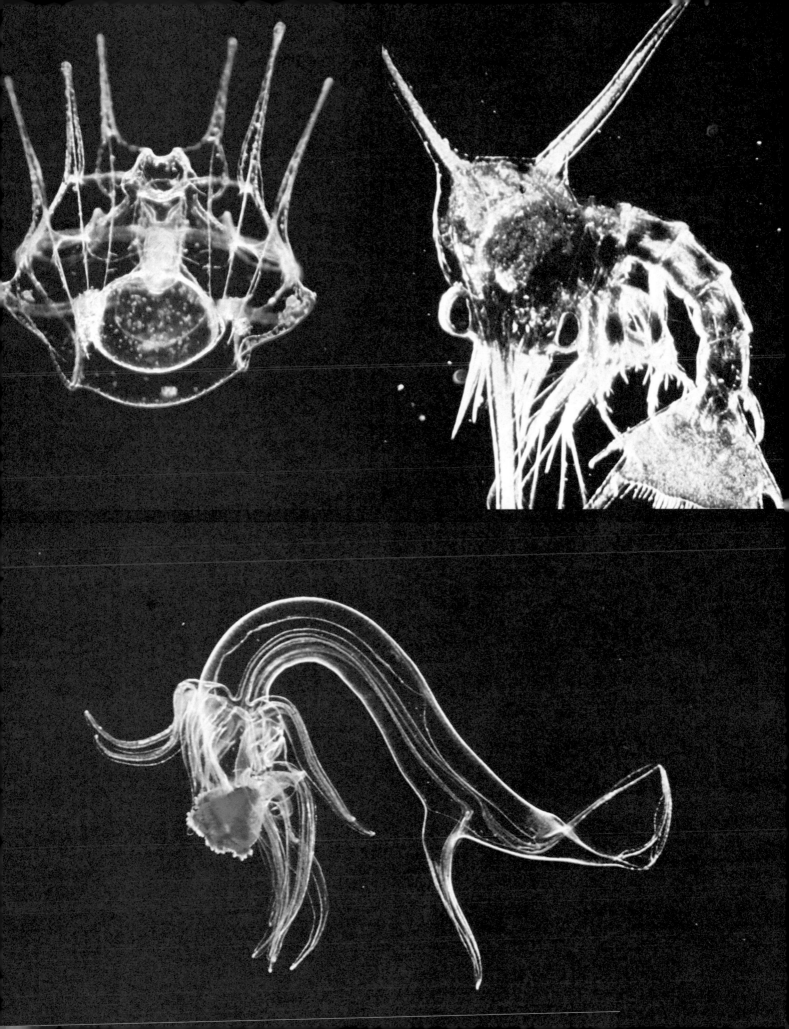

The Act of Life

A vague glow—a sensitive chaos
    of dust and gases—
Shrank into a red earth
    of molten rock and steam
But—since immemorial monsoons
    the earth is blue
The earth is sapphire
    parading in the firmament grit

        A lightning in protoplasm
            inspired the Sea.
        The Sea has coddled love and life
            and life—the foam of the Sea—
        Spilled over the continents.

A lightning in protoplasm
    has triggered trillions of sparks
In spores for the liquid meadows
    in sperm and egg for the night of the squid
For the salmon odyssey
    and for the faithful albatross.

        A lightning in protoplasm
            has launched our comets to the moon
        To Mars to Venus to Jupiter
            and tomorrow will splash our tears to other stars.

The feeble spark that lit our love
    when suns and rocks despair
Life—the water miracle relayed to our mercy—
    Life—in labor and threat from thy latest child.

Index

ILLUSTRATIONS AND CHARTS:

Sy and Dorothea Barlowe—27, 71; Walter Hortens—33, 34, 35; Howard Koslow—43, 77, 90, 91, 104, 105, 109, 124, 125.

PHOTO CREDITS:

Chuck Allen—54; Dr. Herbert R. Axelrod, courtesy TFH Publications Inc.—46–47; Ben Cropp—36–37, 51, 58–59; Jack Drafahl, Brooks Institute—15, 95 (bottom), 138, 139; Gulf Specimen Company/J. J. Rudloe—93; Dr. Walter N. Hess—118–119; Edmund Hobson—56–57, 87; Hyperion Sewerage Plant—92, 140 (right), 141 (top right); William J. Jahoda, from National Audubon Society—11; Tom McHugh, Point Defiance Aquarium, Tacoma, Wash.—55, 112, 133; Marineland of Florida—120–121; Chuck Nicklin—44, 45; V. F. Penfold—96, 113; Photography Unlimited: Ron Church—64 (top and middle), 102; Dr. Kenneth R. H. Read—19, 98–99; © 1967, Fred M. Roberts—40 (bottom), 89; William M. Stephens—63, 97 (top), 100–101, 117; Valerie Taylor 84–85; Joe Thompson—29; Judson Vandevere—130–131; Wards Natural Science Establishment Inc., Rochester, N.Y., and Monterey, Calif.—41 (top right); © Douglas P. Wilson—12–13, 39, 67, 141 (bottom and top left).